Error-correcting Codes

Error-correcting Codes

A mathematical introduction

John Baylis

Deparment of Mathematics

Nottingham-Trent University

UK

CHAPMAN & HALL

London · Weinheim · New York · Tokyo · Melbourne · Madras

Published by Chapman & Hall, an imprint of Thomson Science, 2–6 Boundary Row, London SE1 8HN

Thomson Science, 2–6 boundary Row, London SE1 8HN, UK

Thomson Science, 115 Fifth Avenue, New York, NY 10003, USA

Thomson Science, Suite 750, 400 Market Street, Philadelphia, PA19106, USA

Thomson Science, Pappelallee 3, 69469 Weinheim, Germany

First edition 1998

Thomson Science is a division of International Thomson Publishing [ITP logo]

Typeset in Great Britain by Focal Image Ltd, 20 Conduit Place, London W2 1HZ
Printed in Great Britain by St Edmundsbury Press, Bury St Edmunds, Suffolk

ISBN 0 412 78690 7

A catalogue record for this book is available from the British Library
Library of Congress Catalog Card Number: 97-75115

To Catharine

Contents

Preface

It has been said many times, and that in no way diminishes its truth, that coding theory is a striking example of the applicable power of many diverse branches of mathematics. For students who have been trained in traditional pure mathematics a course in coding theory can provide an icing on the cake. This book is not written for them. It is more for that much larger population of students who find themselves in an education system driven by an 'application' philosophy, which regards pure mathematical foundation as an expensive luxury, in spite of all the accumulated experience that such a system is short-sighted and self- defeating.

My experience of teaching a final year course on coding at The Nottingham Trent University has helped to crystalize the aims of this book. They are: to motivate the need for error-correcting codes; to show some of the ways in which this need is met with codes in actual use; to explain some of the mathematical tools on which codes depend; to show some of the connections which make coding theory such an attractive subject; to get students to the point where they can realistically tackle more advanced work on coding; and finally, I wanted to make the book as self-contained as possible.

The main mathematical underpinning is number theory, linear algebra and polynomial algebra. The first and last of these are all too often in the 'expensive luxury' category, so I have assumed no background in these topics. Students generally meet linear algebra early in their course, then have plenty of time to forget what a vector space is before their final year. For this reason only minimal background is assumed. Although the main reason for all this mathematics is its application to coding, I did not want the material on mathematical topic x to be merely 'x for coding'. Chapter 3, for example, discusses Fermat's theorem not because of any use in this book, but because it, or its analogue in a finite field, is vital in more advanced work on coding, in cryptography (which students may well be learning simultaneously), and in a wide range of computer science topics. A couple of more recreational aspects of number theory are also included under the 'wider education' umbrella.

The principle mathematical topics *excluded* from this book are finite fields and group theory. One can't do everything! These topics could fit nicely into a second volume containing, for example, BCH codes, Reed–Solomon codes, automorphism groups and connections between codes and finite geometries. The fact that Z_p is a finite field is, of course, mentioned – we need multiplicative inverses to do the linear algebra, and a couple of group-theoretic ideas (cosets, closure) crop up implicitly in syndrome decoding, but the detailed machinery of these topics is not needed in this volume.

The exercises form an integral part of the text. The reader gets indications throughout the text that he has reached the point where he should do exercise n. The exercises consolidate the immediately preceding text, prepare the ground for the following text, or sometimes just provide extra background. Readers are strongly advised to do, or at least to read the solutions of the exercises as they come to them.

I have tried to spell things out in rather more detail than is customary in a text at this level, and for this I offer the following explanation, but not apology! Learning from a book is a non-trivial skill which takes time and effort to acquire. In recent years the disincentives to acquiring it have mounted: pressure on student finances and time; the production of cheaper 'home-grown' course-specific notes; the tendency of authors to write books for ideal students. Students, even the good conscientious ones, are usually not ideal, and therefore need all the help they can get. I hope then that readers will find the book 'student-friendly' and will be able to concentrate their efforts on the exercises, not hampered by too many obscurities in the text!

A book of this nature does not, of course, appear out of thin air. Debts to other books will be obvious, but I would like to thank specifically Ray Hill, whose book has been top of the Nottingham Trent University student's reading list for several years, and Oliver Pretzel, whose understanding attitude to our introductory examples being almost identical was a great relief. Chapter 3 owes much to my experience as an Open University tutor. The whole book owes much to The Nottingham Trent University students acting as experimental subjects for much of the material. Its final form is a consequence of the transformation of my handwriting into elegant LaTeXby Anne Naylor and Salma Mohamedali. The friendly gentle negging of Stephanie Harding and Mark Pollard at Chapman & Hall eased the transition from typescript to book. To all concerned I offer sincere thanks.

1

Setting the scene

1.1 The problem

Next Saint Valentine's day you may be lucky enough to receive the message

I LMVE YOU

from someone whose sentiments are pretty clear, in spite of not quite having mastered the word-processor.

This is a trivial example but also a very instructive one. It will lead us to quite a good understanding of what coding theory is all about. Before reading on, think for a few minutes about : how you know the message contains an error; why you are confident about the location and number of errors; why you are confident that you can correct the error(s); whether other similar cases may not be so easy to correct.

Let us now consider those questions in order. Words in English, and in any other natural language, are strings of letters taken from a finite alphabet, but not all such strings correspond to meaningful words: 'I' does; 'Y O U' does; 'L M V E' doesn't; so we have detected an error in the second word. As for the location being letter 2 of word 2, this seems the most plausible thing which could have gone wrong, as the simple remedy of changing that M to an O restores a meaningful word and makes the complete message plausible. No other replacement for the M does this, so we have done something about question 3 too. But before we get too confident, couldn't the correct message have been 'I LIKE YOU' with errors made in the middle two letters of the second word? Well, yes, but you could argue that two errors are less likely than one because the sender is known to be not too bad at word processing. Finally, it is possible to receive a message which contains one error but the recipient can be unaware of its existence. A simple example of this is the received message:

I LOVE LOU

I have no way of knowing whether this is really the sender's intention, or whether the real message is

I LOVE YOU

transmitted with one error. To make progress here I would need to use information other than the message itself, for example, my name is not Lou, and I don't know anyone of that name. In other words, I would be using the context of the message.

It is useful at this stage to make three general points arising from our examples:

1. Natural languages have built-in *redundancy*, and it is this which gives hope of being able to detect and correct errors.
2. Our confidence that our error correction is valid is greater if we have some assurance that a small number of errors is more likely than a larger number.
3. It may happen that we can detect that an error has been made, but cannot correct it with confidence.

The form in which we have already met redundancy is that not all strings of letters are meaningful words. Another is illustrated by observing the effect of simply cutting out large chunks of the original message. For example, every vowel and every space between words has been deleted from an English sentence and the result is

`THPRNCPLXMPLFFNCTNWWSHTMPHSSNDLLSTRTTTHSPNTSTHTFCMPTTNBYCMPTRPRGRM` .

You will probably be able to reassemble the original message without too much difficulty, perhaps with a little help from an intelligent guess about the context.

An example of the third point is a one-word message received as L M V E. You can be sure an error has been made, but, without additional information, will be unable to decide between L O V E and L I V E. [Ex 1]

1.2 The channel – cause of the problem

We have seen how some features of a language, principally its redundancy, and facts about the sender, like being prone to make rare single errors but less likely to make more, can help the receiver to recover a slightly garbled message. If our messages are important enough to be guaranteed error-free could we not simply put the onus on the sender to use a sufficiently thorough system of checks that no message containing errors was ever sent? The only thing which prevents this from being an excellent idea is that in most significant applications the errors do not arise from mistakes on the part of the sender, but rather, as a result of what happens to the message after leaving the sender and before arriving at the receiver. In other words, the *communication channel* has a vital role to play. For instance, in our first example the O received as an M may be the result of smudging.

A more clear-cut example is a conversation between two people at a rather noisy party. The speaker's words could be uttered perfectly clearly but his listener could fail to receive some of them, or receive a distorted version, due to cross-talk from other conversations, or loud music, etc. Here the idea (or fact, or comment, or etc) which the speaker wishes to transmit has to be *encoded* into a form suitable for the *channel.* Here the channel is the air between speaker and listener; the words of the message are encoded as pressure waves in the air; these impinge on the listener's ear; finally a complex *decoding* mechanism reinterprets these waves as words.

In coding theory and elsewhere the everyday term *noise* is used to denote any inherent feature of a communication channel which tends to distort messages, and which are generally outside the control of both sender and receiver. Noise is a fact of life in both the elementary examples we have mentioned so far and in much more sophisticated technological examples. To mention three fairly obvious ones: telephone lines are subject to unavoidable crackle; the signals which carry pictures of remote bits of the solar system back to Earth can be distorted by cosmic rays and solar flares; information stored in computer memories can be corrupted by the impact of stray alpha particles, ... and so on.

1.3 Cunning coding – solution of the problem

First consider a very simple situation in which the sender only needs to send one of two possible messages, say 'yes' or 'no', 'stay' or 'go', 'attack' or 'retreat', etc. These days messages are often sent as digital pulses rather than as written or spoken words, so let us suppose our two possible messages are coded as 0 and 1. This has the virtue of simplicity, but the price to be paid is that a single error can mean disaster. If the general receives intelligence that his troops are vastly outnumbered, so sends '0' meaning 'retreat'and a stray bit of electromagnetic noise corrupts this to '1' for 'attack', then the consequences could be most unpleasant, so unpleasant that one could not be expected to tolerate such occasional errors even if very rare. The source of the problem is that there is no redundancy at all in this system, so no chance of detecting that the received message is an error. It has been said, and I paraphrase slightly, that modern coding theory is all about replacing the redundancy we lose in going from natural (English) to artificial (digital) language in a sufficiently cunning way to enhance the error-correcting capability of the language.

It is very easy to give examples of simple ways of achieving this. If we stick with our primitive two-message system but this time agree to transmit 00 whenever we intend 0 or 'retreat' and 11 whenever we intend 1 or 'attack', then there are just two messages, 0 and 1, and these are encoded by the *codewords* 00 and 11 respectively. Now suppose the channel is subject to noise which can corrupt codewords by changing a 0 to a 1 or a 1 to a 0, and the codeword 00 is sent. One of four things can happen. It may be received

as	00	with no corruption
or	01	with the second digit corrupted
or	10	with the first digit corrupted
or	11	if both digits are corrupted,

and a similar set of possibilities occurs if 11 is sent.

Now put yourself in the position of the receiver who only has available his pair of digits and no knowledge of what was sent. (To make things more friendly we'll make Siân the sender and Rhidian the receiver.)

If Rhidian receives 00 he knows that either 00 was sent with no interference from noise, or that 11 was sent but both digits were corrupted. He is in a similar position if he receives 11. On the other hand if 01 is received it is certain that one of the two digits has been corrupted *because 01 is not a codeword*, but he has no idea whether 00 was sent and the second digit corrupted or 11 was sent and the first digit corrupted. Likewise if 10 is received.

It seems then that so much uncertainty still remains that nothing worthwhile has been achieved. But notice how the situation changes if we know that under no circumstances can both digits be corrupted. Now if 00 or 11 is received Rhidian knows that this was Siân's message. If he gets 01 or 10 he knows there is an error (but still doesn't know where). In practice we can never of course guarantee that it is impossible for both digits to be changed – after all we have no control over the channel. But also in practice we can do something almost as good: to be any use the channel only rarely induces errors so that the probability p that a randomly chosen digit of a randomly chosen message is corrupted is small; the probability that both digits are corrupted is then p^2 – very much smaller. Suppose p is 10^{-2} so that p^2 is 10^{-4}, and Siân and Rhidian have to decide whether to use this channel, with the two-fold repetition code we have described above to send their messages. In the majority of cases messages will survive the noise in the channel and arrive intact, so if Rhidian gets 00 he will be safe in assuming, most of the time, that this was Siân's intension, and similarly for 11. In a small number of cases he will detect that one error has been made, and in very rare cases he will receive 00 (or 11), assume this is correct, and be wrong! To quantify this we shall have to make a couple of assumptions which are realistic for many channels and which are widely used in those branches of coding theory with which this book is principally concerned. These are that:

1. Errors occur at random and independently, so the fact that one digit is corrupted has no bearing on whether or not the next one is;

   and

2. A $0 \to 1$ corruption is just as likely as $1 \to 0$, so the channel is said to be symmetric.

   Now suppose Siân sends 00.

$$\begin{aligned} P(\text{Rhidian receives 00}) &= 0.99 \times 0.99 = 9801 \times 10^{-4} \\ P(\text{He receives 01 or 10}) &= (0.99 \times 0.01) + (0.01 \times 0.99) = 198 \times 10^{-4} \\ P(\text{He receives 11}) &= 0.01 \times 0.01 = 1 \times 10^{-4} \end{aligned}$$

Because of the symmetry of the channel an identical analysis applies when she sends 11.

The decision which our two protagonists now have to make can be based on the following questions:

1. In about 2% of cases Rhidian will detect an error so what are the consequences of just ignoring these messages? Or is it feasible for him to communicate back to Siân a message along the lines 'I've detected an error in your last message. Would you please re-transmit, and let's hope I receive it intact this time'?
2. In just one out of every ten thousand cases Rhidian will misconstrue Siân's intention (and probably act on this false information). Can they live with this, or are the consequences so serious that they need to consider using a better channel, or if they are stuck with the channel devise a better message encoding scheme?

Coding theory is a powerful mathematical tool for dealing with the very last of these points, but of course the final *decision* has to depend on the nature of the messages and the circumstances in which they are sent.

The scheme described above is called a *binary* code, since its *alphabet* (set of available symbols) consists of only two symbols, 0 and 1.

Its *codewords* are those strings of symbols which may be sent, 00 and 11 in our case.

It is a *block code of length 2*, meaning that its codewords are all of the same *length*, the length being just the number of alphabet symbols per codeword.

We shall generally call a sequence of alphabet symbols of the right length (two in this case) just *words* or *strings* or, later, *vectors*. Thus the complete set of words is $\{00, 01, 10, 11\}$.

This code is called a *one-error-detecting* code. What is meant by this is that *in all cases* in which one error per word is made, the receiver can detect this fact. Notice that this does *not* mean the receiver knows *where* in the word the error occurs.

Let us return to English for a moment in order to make some important distinctions. Our word-processing friend sends us DAT, and we know she makes at most one error per word. (We can think of the code as the set of all standard English words, or just three letter words if we want a block code.) DAT is not a codeword so we have *detected* an error.

Table 1.1 *The performance of the binary 3-fold repetition code when the receiver assumes at most one error.*

Code word sent	Word received	Interpretation made by receiver	Correct codeword recorded
000	000	000	yes
	001	000	yes
	010	000	yes
	100	000	yes
	110	111	no
	101	111	no
	011	111	no
	111	111	no

We are incapable of *locating* it for the correct codeword could have been SAT, DOT, or DAB, and even if we had extra information to the effect that the error was definitely in the first letter, this would not enable us to *correct* the error as the codeword could have been BAT, CAT, EAT, FAT, ...

Note that binary codes have the nice property that error location is equivalent to error correction because if you know there is an error in the ith place you simply change the symbol in that place to the other one!

Now return to our binary code and imagine the following scenario. Messages from Siân are important and should not be ignored! Moreover, the communication line from Siân to Rhidian is only one way, so what is Rhidian to do when he detects an error? Well, Siân can modify her coding scheme and use instead a three-fold repetition code. That is, she uses 000 and 111 as her two codewords, and they still consider that two or more errors per word happen sufficiently infrequently not to matter too much. So now Rhidian is going to interpret all received words as if at most one error has been made. How well he performs is shown in Table 1.1 below, in which we have taken 000 to be the codeword Siân sends.

In this case it is fairly clear how Rhidian makes his interpretations. For example, looking at line 5, 110 is received. This is not a codeword so there must be at least one error. On the assumption of at most one error the interpretation has to be 111, an error having been made in the third digit, for the alternative interpretation as 000 would involve two errors (in digits one and two). By assuming at most one error Rhidian has of course made the wrong interpretation. However, a glance at the table shows that the correct interpretation is made in all cases of one or no error, so we describe this code as one-error-correcting.

Another piece of salient advice to theoretical coding theorists is that a code is, in practice, only as good as its decoding algorithm. If you have by means of some clever mathematics invented a code capable of correcting lots of errors the whole project could be rendered useless if the calculations required to do the decoding are beyond the reach of current computing technology, or would take so long that the messages would only be decoded long after they had ceased to be relevent! It has to be admitted that much research in coding theory pays no attention to such practicalities. The reason is that the researchers themselves are pure mathematicians, for whom the beauty of the mathematics involved in inventing and investigating the properties of codes is sufficient unto itself. Now is an appropriate time for an admission from me. The motivation for writing this book is mainly that coding theory is such a brilliant display cabinet for so many gems of mathematics, so the content of the book was chosen mainly on aesthetic rather than practical grounds. The practicalities which are discussed are also chosen for their nice mathematical features. I hope that statement will discourage readers from writing to me with threats of invoking a Trades Description Act for mathematical books, but please do write about anything else!

Fortunately the code under discussion has a very simple decoding algorithm which can be described precisely as in Frame 1.1.

Frame 1.1 The decoding algorithm for the binary 3-fold repetition code used in error-correcting mode.

1. Count the number of occurrences of each symbol in the received word.
2. Decode it as xxx where x is the symbol which occurs with the greater frequency.

Even if Siân and Rhidian have a two-way communication channel there are still good reasons why they may prefer to seek an error-correcting code rather than one which is only error-detecting. If messages need to be acted upon quickly there may be no time to send a message back requesting retransmission. Or if Siân is a remote camera on an artificial satellite orbiting Jupiter, by the time Rhidian (the Space Control Centre) has received a blurred image and requested a retransmission, Siân has disappeared round the other side of the planet so can't oblige!

We have seen that the three-fold repetition code will function as a one-error-correcting code, but if error detection is considered good enough it can be used as a *two*-error-*detecting* code as follows. Look at Table 1.1 again, and follow the instructions given in Frame 1.2. Note that the only circumstances in which this scheme can fail is when three errors are made so that 000 is sent but 111 is received and accepted as the correct word, or vice-versa. All instances of two or fewer errors are detected.

Frame 1.2 The decoding algorithm for the binary 3-fold repetition code used in error-detecting mode.

1. If 000 or 111 is received, accept this as the transmitted code word.
2. If anything else is received declare an error in this word and request retransmission.

Another interesting and important feature of this code is that you cannot hedge your bets. You have to decide in advance whether to use the code in error-correcting or in error-detecting mode, for in the former receiving 010 would trigger the response of deducing that the intended codeword was 000, but in the latter the response would be to ask for retransmission.

Before leaving this code let us quantify its rather impressive performance when used in the error-correction mode. Again we denote by p the symbol error probability, and compare performances with and without coding. Using the code,

$$\begin{aligned} & P(\text{received word wrongly interpreted}) \\ = \; & P(\text{channel induces two or more errors}) \\ = \; & P(2\text{ errors}) + P(3\text{ errors}) \\ = \; & 3p^2(1-p) + p^3 = \; p^2(3-2p) \end{aligned}$$

and when $p = 10^{-2}$ this is 3×10^{-4} approximately. In other words, only about three messages out of every ten thousand would be decoded wrongly.

Without any coding (just plain 0 or 1 is sent) one in every hundred messages are wrong.

Like many things in real life there is a price to be paid for this improvement. One of them is that we have to send three symbols for every one we want to get across, and this *three-fold message expansion* means that the time taken (and probably the cost too) to send our messages is three times what it would have been had we dispensed with the advantage of coding.

This brings us sharply up against one of the main problems of coding theory: how can the redundancy necessary to achieve good error correction and/or detection be arranged so as to minimize the message expansion? One of the earliest good answers to this problem is described in the next chapter. [Ex 2, 3]

1.4 Exercises for Chapter 1

1. Try to recover the message with vowels and spaces omitted at the end of section 1.1.
2. The binary 3-fold repetition code is to be used for one of two possible channels. The first is a non-symmetric channel which induces $1 \to 0$

errors with probability α and $0 \to 1$ errors with probability β. The second is a symmetric channel with an overall symbol error probability equivalent to that of the first channel, $\frac{1}{2}(\alpha+\beta)$. The channel is to be used in 1-error-correcting mode. If $P(sym)$ and $P(non-sym)$ denote the probabilities that a word is wrongly decoded using the respective channels, show that, provided 000 and 111 are equally likely to be sent,

$$P(sym) - P(non-sym) = \frac{3}{4}(\alpha-\beta)^2(\alpha+\beta-1).$$

Hence decide what other information is needed in order to decide which channel to use.

3. The symbol error probability of a symmetric channel is p. The messages are all binary of length 3 and these are encoded by adding a fourth bit to each message so that the total number of ones in each codeword is even. What proportion of received words will contain undetected errors?

2

Reducing the price

2.1 Hamming's solution

In the last chapter the problem of finding a method of coding which would correctly retrieve the transmitted codeword whenever it was received with a single error was solved. But the solution, the three-fold repetition code, was rather unsatisfactory because of the associated three-fold message expansion. So the question now is whether we can find a code with smaller message expansion but with equally good error correcting capability.

In 1948 Richard Hamming, working at the Bell Telephone Laboratories, discovered a technique for doing just this, and Hamming codes are widely used to protect computer memories against failure. We shall see exactly how this is done later, but for the moment we concentrate on describing one of the simplest of the Hamming family of codes. It is another *binary* code so again we have an alphabet {0,1} of size two, and the set of messages which can be sent is the set of all binary strings of length four. How these are encoded ready for transmission is best described by referring to Figure 2.1. It shows three circular regions A, B and C inside a rectangle R, arranged to divide the rectangle into eight areas which are labelled 1–8.

Figure 2.1(b) shows how a particular message, 0100, is encoded. First, the four 'bits' (binary digits) of the message are placed, in order, in regions 1 to 4. The redundancy is added as extra bits 5, 6 and 7 according to a simple rule: the total number of 1s in each of the circles A, B and C must be even. Since the first four bits fill three of the four regions into which A, B and C are split, the bits to go in regions 5, 6 and 7 are uniquely determined by this rule. In this case our message 0100 gets encoded as 0100011. Notice that the message expansion factor is now only 1.75 instead of 3, so this is a vast improvement provided we can still correct all instances of single errors. This is indeed the case, and you are invited to prove this in one of the exercises for this chapter. First we work through a couple of examples of the decoding process. [Ex 1]

Suppose Rhidian gets 0101010. He first puts the received bits, in the right order, back into the regions 1 to 7 of the diagram. (see Figure 2.2). Then he does the *parity checks* on each of A, B and C. That is, he just records whether each of these areas contains an even number of ones (√) or an odd number (×). In this case all three fail so he knows that at least one error has occurred, because the transmitted word was deliberately devised to have all three even.

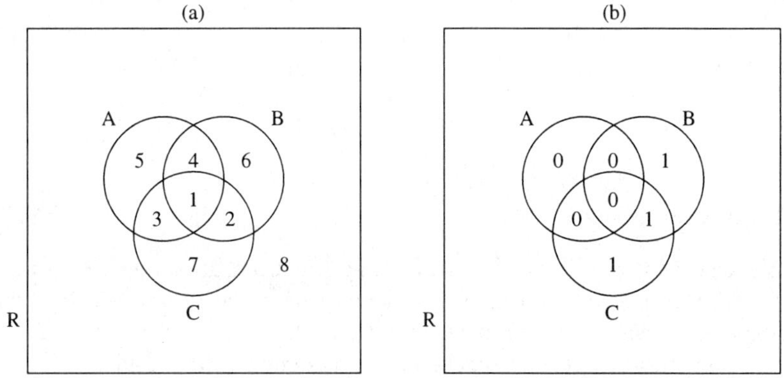

The numbering of the regions

0100 is encoded as 0100011.

Figure 2.1 *The Hamming (7,4) code.*

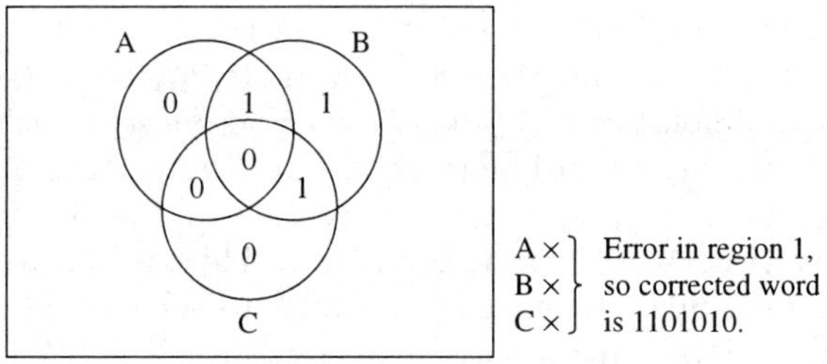

Figure 2.2 *Decoding the received word 0101010.*

He can actually do better than this by the following reasoning: if there is only one error it has clearly affected all of the regions A, B, C, and there is only one bit which is in all three of A, B, C – the first one. Hence the error is in bit one, and the correct word must therefore be 1101010.

To take one more example, suppose 1100101 is the received word. You should check that this time the parity check works for A and B but fails for C. Hence the error only affects C, and the only bit which does this is bit 7. So this is the corrupted bit and the corrected word is 1100100.

The importance of this example is in showing that it is just as easy to correct an error in one of the redundancy bits (5, 6, 7), as in one of the message bits (1, 2, 3, 4). This is just as well, because if the channel noise is a stray cosmic ray it has no interest in whether it hits one of the first four or one of the last three bits!

The code we have just described is sometimes called, for obvious reasons, the binary Hamming (7,4) code. Frame 2.1 makes its decoding algorithm explicit. [Ex 2, 3]

Frame 2.1 Decoding algorithm for the binary Hamming (7,4) code, assuming at most one error.

1. Put the bits of the received word, in order, in regions 1 to 7.
2. Do the parity checks on A, B and C.

   If all are correct accept the received word.

   If the check on just one of A, B, C fails, the error is in bit 5, 6 and 7 respectively.

   If B,C or A,C, or A,B fail (but the check on A,B,C respectively is correct), the error is in bit 2, 3 or 4 respectively.

   If all three fail, the error is in bit 1.

2.2 Can anything be done if two errors occur?

Suppose that Siân decides to send 1100. She encodes this (correctly) as 1100100, but during transmission bits 2 and 6 are corrupted so that Rhidian receives 1000110. If you put this back into the decoding diagram and decode *on the assumption that there is at most one error*, you can check that the outcome is to declare an error in bit 7. Notice that this is not one of the places where there is actually an error, so the process which works beautifully when there is only one error can confuse the situation still further if there are two. So our advice to Siân would have to be: 'If you are sure the channel cannot induce two or more errors per word, or if the probability of this happening is so small that you can live with the corresponding small proportion of your messages being misinterpreted, then use the Hamming (7,4) code. If not, look for a different code'.

One of the different codes Siân may consider involves only a small adaptation of the (7,4) code. You may have wondered why the exterior region of Figure 2.1(a) received the label '8'. We are about to use it. We encode the 4-bit messages as 8-bit code words this time to obtain the binary Hamming (8,4) code. Bits 5, 6 and 7 are determined in the same way as before and bit 8 is an *overall parity check bit.* That is, it is chosen to make the total number of ones in the whole 8-bit code word even. If Siân uses this system, instead of sending 1100100 she will send 11001001. Sticking to our example of errors in bits 2 and 6, Rhidian receives 10001101, which he dutifully puts into the decoding diagram – refer to Figure 2.3. Initially Rhidian has an open mind about the errors and begins to narrow down his options in the light of the checks: A and B work, C fails and the overall check works. The fact that any of the checks fail tells him there must be at least one error. How many? Notice that if you start with any binary string, each time one error is made the overall parity must change from odd to even or even to odd. The transmitted word has even overall parity because of the way the (8,4) code is defined, so the fact that the received word is still even must mean that an even number of errors have been made.

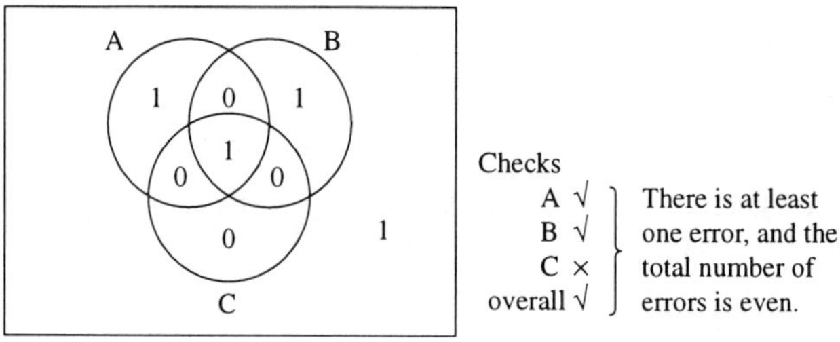

Figure 2.3 *Using Hamming (8,4) to interpret 10001101.*

If the practical situation is such that the possibility of three or more errors can be ignored safely, then Rhidian has *detected* that there are two errors in this word. Where could they be located? The first clue to use is the fact that the check on C failed so exactly one of the errors is in C, that is, in region 1, 2, 3 or 7. If it is in 1 and we make the correction in region 1 this will correct the parity of C, but make A and B odd, so the remaining error must lie in the region which affects A and B but not C, that is region 4.

If the C error is in 2, changing this bit will make C even, leave A even, but make B odd, so the remaining error must lie in the region which affects only B, that is region 6.

By continuing this line of reasoning for the other two cases you should see that the four possibilities for the location of the errors are (1,4), (2,6), (3,5) or (7,8). The outcome then is that although Rhidian has not been able to correct the two errors with certainty he has been able to do better than pure detection in that he has narrowed down to four the possible locations. In the exercises we ask you to check by similar arguments that whatever the transmitted word, and wherever two errors are made, the Hamming (8,4) code enables you to detect this fact and to reduce to four the set of possible error locations. In anticipation of this, Frame 2.2.2 presents a decoding procedure for this system. [Ex 4, 5, 6, 7]

Frame 2.2 Decoding algorithm for the binary Hamming (8,4) code, assuming at most two errors.

1. Put the bits of the received word, in order, in regions 1 to 8.
2. Do parity checks on A,B,C and the overall check.

   If all are correct accept the received word.

   If the overall check fails and at least one of the other checks fails, deduce that there is only one error, that it is in bits 1 to 7, and correct it as for the Hamming (7,4) code.

   If the overall check is the only one which fails, deduce that there is one error, that it is in bit 8, so make the correction.

   If the overall check works but at least one of the others fails, declare that there are two errors and request retransmission.

Both the (8,4) code and the three-fold repetition code considered in Chapter 1 can be used to correct single errors and detect up to two errors per word. But the Hamming code wins when they are compared for message expansion (3 for the repetition code but only 2 for the (8,4) code). The Hamming code has another advantage: you do not have to decide in advance whether to use it in error detection or error correction mode. The reason is that in the repetition code the reception of a non-codeword can indicate either an error in one of the bits or errors in the other two bits, and there is no way of telling which has occurred. But in the (8,4) code if one or more of the checks on A, B and C fails, the overall check will always distinguish between one error and two errors.

2.3 An alternative use of Hamming codes – erasures

In certain channels it is possible for bits to be wiped out or rendered unrecognizable rather than corrupted from one alphabet symbol to another. Such a fault is called an *erasure*, and the codes we have discussed up to this point can deal with these too. Apart from a brief investigation in chapter 5, erasures will play no significant part in this book so we confine ourselves to one example here, and refer you to Exercises 8 to 12 to pursue this further and get some more experience with our diagrammatic representation of Hamming codes.

Consider a channel in which erasures can occur but not $0 \leftrightarrow 1$ corruptions, and suppose words are transmitted using the (7,4) code, and one word is received with the second and fourth bits unrecognizable. Follow the procedure shown in Figure 2.4, of starting to decode in the usual way. Represent the 'smudged' bits as x and y, then the parity checks tell you that the transmitted codeword is one in which x and y have to satisfy the conditions that $y+2, x+y+1$ and $x+3$ must all be even. Since x and y

can only be 0 or 1 it is easy to see that there is only one possible solution: $x = 1, y = 0$, and the code word is 1110001.

Notice that at least in this example the (7,4) code has managed to survive two *erasures*, whereas we know that it cannot cope with two *errors*. This is to be expected as in the case of an erasure the location of the trouble is known. The exercises investigate how representative this example is of the general situation. [Ex 8 – 11]

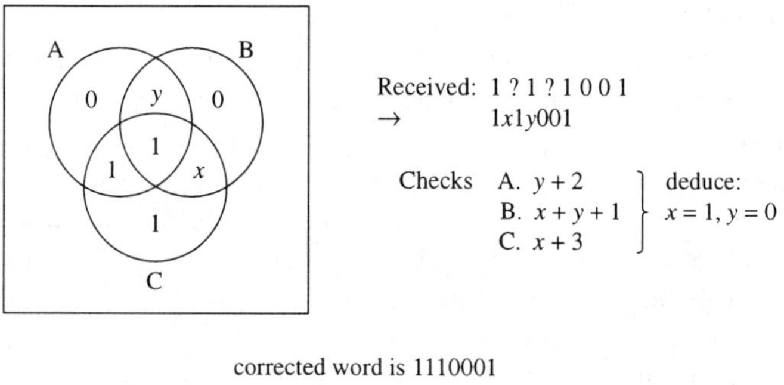

Figure 2.4 *Interpreting two erasures with the (7,4) code.*

2.4 What really makes a code work? – Hamming distance

Why is the Hamming (7,4) code so good at correcting single errors? The answer depends on the following claim – that *any two 7-bit codewords differ in at least three places.* I ask you to take this on trust for the moment while we look at its important consequence. Suppose for example that the transmitted word w_t is $x_1x_2x_3x_4x_5x_6x_7$ and that noise changes the sixth bit, so the received word w_r is $x_1x_2x_3x_4x_5x_6'x_7$, where $x_6 \neq x_6'$. Now let w be any codeword other than w_t. It differs, according to my claim above, from w_t in at least three places. If one of these places is the sixth, then w differs from w_r in at least two places, and if not, in at least four places. To summarize, if only one error is made, so that the received word differs from the transmitted word in only one place, it differs from every other codeword in at least two places. Our diagrammatic decoding method operates by interpreting the received word as that codeword which can be obtained from it by changing at most one bit, and we have shown (subject to the claim) that there is only one codeword with this property. [Ex 12]

So now it remains to substantiate the claim! A direct but very tedious method would be to list all the codewords and for each pair count the number of places at which they differ, but this would involve doing $\binom{16}{2} =$

120 counts. We can reduce the work (and enhance insight!) if we make use of the symmetry inherent in the code's construction.

If the diagram for the code is not yet fixed in your memory you will need to refer to Figure 2.1(a) again. We first classify the bits of a word as follows:

$$\{x_5, x_6, x_7\} \quad - \quad \text{type 1}$$
$$\{x_2, x_3, x_4\} \quad - \quad \text{type 2}$$
$$\{x_1\} \quad - \quad \text{type 3}$$

where the type number is the number of parity checks affected if a bit of that type is changed. For example, changing x_4 will affect the checks on A and B.

So if we start with a codeword (so that all three parity checks work) and change one bit, then at least one check will fail. Similarly, if two bits are changed you can check that the results are as tabulated below.

Changed bit types	Result of changes
2 and 3	one check fails
1 and 3	two checks fail
1 and 2	one fails (e.g. changex_7 and x_3) or all three fail (e.g. change x_7 and x_4).

Hence two *codewords* (that is, words for which no parity check fails) must differ in at least three places.

The concept we have been using is so important throughout coding theory that it is given a title:

Definition 2.1 If w and w' are words of the same length, over the same alphabet, the Hamming distance between them is the number of places at which they differ, and this number is denoted by $d(w, w')$.

The word 'code' is usually taken to mean the set of all codewords, so the repetition codes of Chapter 1 are $\{00, 11\}$ and $\{000, 111\}$ respectively, and the Hamming (7,4) code is a 16 word code.

You have seen how the error-correcting capability of the Hamming (7,4) code is related to the fact that $d(v, w) \geq 3$ for any two distinct codewords v and w. Shortly we shall prove a couple of theorems which make this connection precise for codes in general, and in anticipation of this we make another definition.

Definition 2.2 If C is a code which contains a pair of codewords whose Hamming distance is δ and there is no pair of distinct codewords whose distance is less than δ, then δ is called the *minimum distance* of C and is denoted by $d(C)$.

It would be perverse to use the term 'distance' in the context of codes unless the Hamming distance bears some similarity to what we understand by distance in ordinary (geometric) language, where distance is clearly a numerical measure of the separation between two points. It is positive unless the two points happen to be the same, and clearly Hamming distance has the same property (with points replaced by words). More fundamentally, there is another feature of distance in Euclidean geometry which is often expressed as 'the shortest path between two points is a straight line'. Or to put it another way, the direct journey from A to B is no longer than the journey via any other point P; $d(A,B) \leq d(A,P) + d(P,B)$. (Refer to Figure 2.5 which makes it plain why this fact is called *the triangle inequality.*) In the exercises you are asked to show that this holds for Hamming distance too. [Ex 13–15]

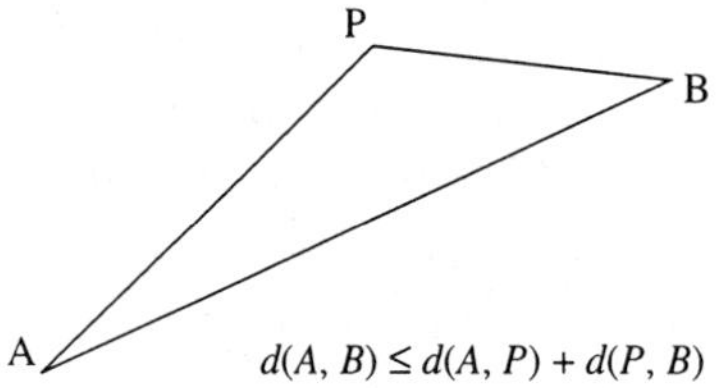

Figure 2.5 *The triangle inequality.*

Hamming distance is but one example of many significant functions sharing these fundamental properties. We digress to mention a couple of examples but if you prefer to concentrate on the coding skip to Theorem 2.1.

First, the geometry can be extended to n dimensions in which points (or vectors) have n co-ordinates, $x_1, x_2, \ldots, x_n$ and distance is defined by a natural generalization of Pythagoras' theorem, namely

$$d((x_1, \ldots, x_n), (y_1, \ldots, y_n)) = \sqrt{[(x_1 - y_1)^2 + (x_2 - y_2)^2 + \ldots + (x_n - y_n)^2]}$$

This distance function or metric is useful to numerical analysts in investigations of the accuracy of methods of solving sets of linear equations in n variables.

Second, a subject called functional analysis uses various measures of separation between pairs of continuous functions defined between two fixed values a and b. Figure 2.6 illustrates two such measures: d_1 is useful if what is important is the worst deviation between the two functions, whereas d_2 is more of an average deviation between them over the whole of the range.

What these and the many other examples have in common is that whether A, B, C are points or words or functions or $\ldots$, d satisfies:

(a) $d(A,B) \geq 0$ for all A, B;

(b) $d(A,B) = 0$ if and only if $A = B$;

(c) $d(A,B) = d(B,A)$ for all A, B;

(d) $d(A,B) \leq d(A,C) + d(C,B)$ for all A, B, C.

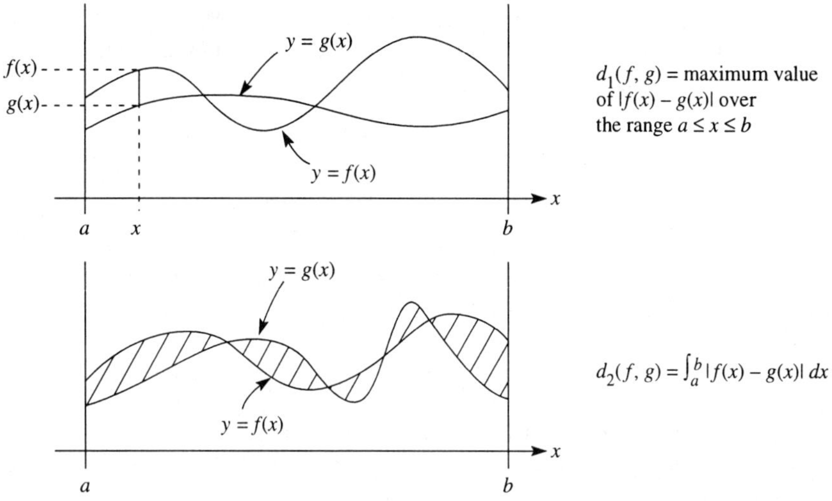

Figure 2.6 *Two 'distances' defined on the set of functions continuous on the interval [a, b].*

Indeed, there is a general theory of *metric spaces*, which are just sets of objects endowed with a function d which satisfies these four rules.

So you now know that the set of words of fixed length n, over any fixed alphabet, is a metric space in which d is the Hamming distance.

We now come to the results which explain the significance of the minimum distance of a code. C denotes any block code over any alphabet. Recall that error-detection means observing that the received word is not the transmitted word, and we assume that error-correction is done by decoding the received word to the codeword at smallest Hamming distance from it. If there is more than one such codeword this fails, and it also fails of course if the codeword 'nearest' to the received word happens not to be the transmitted word.

Theorem 2.1 $d(C) \geq \delta$ if and only if C is $\delta - 1$ error-detecting.

Proof. Let $d(C)$ be at least δ. If any codeword has at least one but fewer than $\delta - 1$ of its digits changed, the result cannot be another codeword.

Conversely, suppose $d(C) = \gamma < \delta$. Then C contains codewords x and y for which $d(x, y) = \gamma$. So if x is sent and the channel induces errors which result in y being received, then the receiver will assume y was sent. Only γ errors have occurred (and $\gamma \leq \delta - 1$) so the code is not $\delta - 1$ error-detecting.

□

Theorem 2.2 $d(C) \geq 2\varepsilon + 1$ if and only if C is ε error-correcting.

Proof. First suppose $d(C) \geq 2\varepsilon+1$. If codeword x is sent and u is received, at most ε errors having been made, we wish to show that u is closer to x than to any other codeword y. To do this, we have

$$\begin{array}{rcll} 2\varepsilon + 1 & \leq & d(x,y) & \text{by the definition of } d(C) \\ & \leq & d(x,u) + d(u,y) & \text{by the triangle inequality} \\ & \leq & \varepsilon + d(u,y) & \text{since } d(x,u) \text{ is at most } \varepsilon \end{array}$$

Hence $2\varepsilon + 1 \leq \varepsilon + d(u,y)$, so $d(u,y) \geq \varepsilon + 1$ as required.

Conversely, suppose $d(C) < 2\varepsilon + 1$ and show that a word can suffer ε or fewer errors but still fail to be correctly decoded: let x and y be codewords such that

$$d(x,y) = d(C) = \alpha < 2\varepsilon + 1$$

If α is even define the word u as follows: take x and choose $\frac{\alpha}{2}$ of the places where it differs from y; change the symbols in these places to the corresponding symbols of y. If α is odd do the same with $\frac{\alpha+1}{2}$ places.

Now let x be the transmitted codeword and u the received word. The number of errors made in this transmission is $\frac{\alpha}{2} < \varepsilon + \frac{1}{2}$ or $\frac{\alpha+1}{2} < \varepsilon + 1$, so because the number of errors is an integer, it is necessarily at most ε. If α is even $d(x,u) = d(y,u) = \frac{\alpha}{2}$ so the transmitted word x is certainly not the *unique* codeword closest to u, and the decoder is therefore 'confused'.

If α is odd then $d(x,u) = \frac{\alpha+1}{2}$ but $d(y,u) = \frac{\alpha-1}{2}$, so $d(y,u) < d(x,u)$ and the decoder would certainly not decode u to the correct codeword x. □

[Ex 16, 17]

All that we have done up to this point has depended on the belief that interpreting the received word as the codeword which is closest to it (in the sense of Hamming distance) is a sensible strategy. This strategy is called *nearest neighbour decoding*, and we now compare this with an alternative strategy which we would naturally use if we had never heard of Hamming distance. This is *maximum likelihood decoding* and is specified as follows: on getting the received word w calculate, for each codeword c, the probability that c was sent given that w is received. Then decode w to the codeword which maximizes this probability. This was the strategy which guided us, somewhat informally, in our discussion of the Valentine's Day message of Chapter 1.

To compare the two strategies, suppose the code is of length n, w is the received word and d is its Hamming distance from the code word c. Then Prob (c was sent, given w is received) $= p^d(1-p)^{n-d} = \left(\frac{p}{1-p}\right)^d.(1-p)^n$, where p is the symbol error probability.

Now if $p < \frac{1}{2}$, then $\left(\frac{p}{1-p}\right) < 1$, and $(1-p)^n$ does not depend on d, so $\left(\frac{p}{1-p}\right)^d(1-p)^n$ is a decreasing function of d. So choosing c to minimize d corresponds to maximizing [Prob(c sent $|w$ received)]. In other words nearest neighbour and maximum likelihood decoding are equivalent provided $p < \frac{1}{2}$. This condition on p is not at all restrictive since no-one would consider using a channel in which symbols were more likely to be corrupted than not! [Ex 18, 19]

To end the chapter on a light note we describe a party trick based on the Hamming (7,4) code. It was first shown to me by Ray Hill [1].

It is a 'number guessing' game for two players, Gwen the great guesser and Llew the limited liar. Gwen asks Llew to choose a number between zero and fifteen inclusive but to keep it to himself. She then asks him seven questions for which he has a further choice: either answer them all truthfully or answer six correctly but lie in reply to the other. He does not have to reveal whether he is lying at any point. From the answers Gwen is able to say what the number is, whether he has lied, and if so, in reply to which question.

The questions are

Q1. Is the number 8, 9, 10, 11, 12, 13, 14 or 15 ?
Q2. Is the number 4, 5, 6, 7, 12, 13, 14 or 15 ?
Q3. Is the number 2, 3, 6, 7, 10, 11, 14 or 15 ?
Q4. Is the number 1, 3, 5, 7, 9, 11, 13 or 15 ?
Q5. Is the number 1, 2, 5, 6, 8, 11, 12 or 15 ?
Q6. Is the number 1, 3, 4, 6, 8, 10, 13 or 15 ?
Q7. Is the number 2, 3, 4, 5, 8, 9, 14 or 15 ?

To explain how the answers to these reveal all to Gwen, recall the diagram for the Hamming (7,4) code. Express the 'messages' – the numbers 0 to 15 – as four bit strings as below

0	1	2	3	4	5	6	7
0000	0001	0010	0011	0100	0101	0110	0111
8	9	10	11	12	13	14	15
1000	1001	1010	1011	1100	1101	1110	1111

Imagine the four bits placed in order in the regions 1 to 4 of the Hamming diagram. Finally add the bits in places 5, 6, 7 in the usual way, so that Gwen's attempt to guess Llew's number is equivalent to guessing its 7-bit encoding as shown below

0	1	2	3	4	5	6	7
0000000	0001110	0010101	0011011	0100011	0101101	0110110	0111000
8	9	10	11	12	13	14	15
1000111	1001001	1010010	1011100	1100100	1101010	1110001	1111111

Then you may check that each of the seven questions is equivalent to asking whether there is a 1 in a certain region of the diagram. In fact, the scheme is arranged so that question Qn is equivalent to 'Is there a 1 in region n of the diagram?'

Now the connection between the game and Hamming decoding can emerge: Gwen gets in reply to her questions a sequence of seven Ys (yes) and Ns (no), for example NYNNYYN, which she then translates as the 'codeword' 0100110. To 'decode' she has to do the parity checks on regions A, B and C. In practice, if she is aiming to be impressive and do her mind reading without assistance from computing power she has to add bits 1, 3, 4, 5 for A, 1, 2, 4, 6 for B and 1, 2, 3, 7 for C. In this case A and C fail but B works, which corresponds to an 'error' in region 3, so the 'transmitted word' was 0110110, and the first four 'message bits' 0110 correspond to 6, so she is able to tell Llew that 6 was the number he thought of, but he attempted to deceive her at question 3! [Ex 20]

If lying is not allowed, then it is easy to ask just four questions which will do the trick. First ask whether the number is in the first half (0 – 7). If yes, is it in the first half of that (0 – 3)? or if not, is it in (8 – 11)? ...and so on. At each stage the range in which the number may lie is halved, so four questions must yield the answer. But of course this strategy still involves feedback from Llew. Gwen's next question depends on Llew's previous reply, but in the Hamming lying game the questions can be declared in advance. [Ex 21]

2.5 Further reading

Two articles published in the *Scientific American* are useful background reading at this stage. Peterson [2] contains some early (pre-computer) error detection devices, a brief account of how hardware known as a shift register can be used to implement some coding schemes, and some of the unsolved problems of coding theory.

McEliece [3] gives a clear explanation of how a real computer memory chip works and how Hamming codes can be used to protect the memory storage against corruption by radiation. In this application the channel is a temporal one rather than spatial. Information is stored on the chip in order to be read at a later date, and this temporal gap between encoding the message and reading it plays an exactly analogous rôle to the spatial gap between sender and receiver in other applications. The coding principles are of course identical in the two cases.

2.6 Exercises for Chapter 2

1. Encode the messages 1011, 1111 and 0111 using the Hamming (7,4) code.
2. A child's arithmetic homework is sent encoded as a binary string according to the Hamming (7,4) system. The 'messages' are defined be-

low. Encode the questions: 13 + 49, 259 ÷ 7, and decode and answer if possible the three questions:

000111000000100011100100011111000010001110 01

000111100100101101011100011001001110101011111 11

Do you suspect any uncorrectable errors?

'Message'	0	1	2	3	4	5	6	7
4-bit string	0000	0001	0010	0011	0100	0101	0110	0111
'Message'	8	9	+	−	×	÷	space	
4-bit string	1000	1001	1010	1011	1100	1101	1110	

3. Can you prove (preferably not by listing all possible single errors in all sixteen codewords!) that the Hamming (7,4) code will correct all instances of a single error?
4. Using the (8,4) code, encode 1000, 1110, 0011.
5. Decode if possible 11101111, 11010100, 10011100, assuming each word has at most two errors.
6. In the word of the previous exercise found to have two errors, find all the possible error locations.
7. Prove that on the assumption of at most two errors, whenever Hamming (8,4) decoding detects two errors there are always four possible pairs of locations of the errors.
8. The Hamming (7,4) code is used for a channel prone to erasures but not errors. If ?0?0111 is received what was the transmitted word?

   Show that if ??11001 is received, the 'no errors' assumption cannot be valid. Can the correct word be recovered if the assumption is amended to 'at most one error in the recognizable bits'?
9. Show that every word received via the (7,4) channel with two erasures and no errors is uniquely recoverable.

10.(a) Are there any received words with three erasures which the (7,4) code can cope with?

(b) Are there any for which it fails?

11. Does your answer to Exercise 10 change if the (8,4) code is used?
12. How many binary strings of length 7 are there?
    How many of these are codewords of the Hamming (7,4) code?
13. In the geometric triangle inequality, why do we have $\leq$ and not $<$?
14. Prove the triangle inequality for words. That is, if u, v, w are any three words of a code, then

$$d(u,w) \leq d(u,v) + d(v,w).$$

15. How many strings of length n are there if the alphabet has q symbols?
16. What is $d(C)$ if C is: (a) 3 error-detecting; (b) 3 error-correcting?
17. C can be used as an α error-detecting code or as a β error-correcting code. What are α and β if: (a) $d(C) = 4$; (b) $d(C) = 5$; (c) $d(C) = 6$?
18. A ternary code (one whose alphabet size is three) C is

$$\{cbaaa, bcabc, bacbc, aabbc, acccb, cbbab\}.$$

Verify that $d(C) = 2$, so that by Theorem 2.2 C is not 1 error-correcting. However, this only means that nearest neighbour decoding will not correct *all* instances of words received with one error. Find examples of words received with one error which: (a) are correctly decoded; (b) are incorrectly decoded.

19. $C = \{011000, 110110, 000011, 101101\}$. Use nearest neighbour decoding to decode, if possible, the following received words: (a) 010110; (b) 101101; (c) 110011.
20. Practise the lying game sufficiently to become a professional magician next Christmas!
21. If lying is not allowed is it possible to guess the number with four questions which are independent of the replies received?

3

Number theory – arithmetic for codes

3.1 Why number theory?

The main outcome of the previous chapter was the explicit connection between the minimum distance of a code and its error-correcting and error-detecting capability (Theorems 2.1 and 2.2). So a code which is good at correcting errors should have a large minimum distance. Since codes with several thousand codewords are often required the job of designing such a code is daunting, and trial-and-error is really a non-starter. As always, mathematics comes to the rescue, for if we impose some mathematical structure on codes their properties are rather easier to sort out, and there is more hope of devising a feasible decoding procedure – that is, one which is not too expensive and which doesn't take too long.

Most of the codes we shall discuss are *linear* codes, and to understand their significance you need to learn (or revise?) a little number theory and linear algebra, which are the topics of this and later chapters.

The Hamming (7,4) code is one which has this nice linear structure, and to appreciate what this means consider the following experiment. Take any two codewords and write them down, one lined up vertically below the other. I chose

$$\begin{array}{rl} & 0011011 \\ \text{and} & 1000111, \end{array}$$

but I suggest you try your own pair. Now write down another word whose digits are chosen as follows: in each place where the digits of the two original words agree put a 0, and where they differ put a 1. So I would get the result 1011100, and observe that this is not just any old 7-bit string but is another *codeword*. You should observe the same phenomenon with your choice, and Exercise 1 invites you to check that this property holds for all choices of the first two codewords.

There are a couple of alternative ways of thinking out what we have just done. One is that we have carried out an addition sum in the binary system but have ignored the 'carry' digits. Another is that in each column of the addition we have recorded just the remainder on dividing the real sum by two. Thus in this strange addition,

$$\begin{array}{llllll} & 0+1 & = & 1+0 & = & 1 \quad \text{because } 1 \div 2 \text{ is } 0 \text{ with remainder } 1 \\ \text{and} & & & 0+0 & = & 0 \quad \text{because } 0 \div 2 \text{ is } 0 \text{ with remainder } 0 \\ \text{and} & & & 1+1 & = & 0 \quad \text{because } 2 \div 2 \text{ is } 1 \text{ with remainder } 0 \end{array}$$

This process is called addition modulo 2, often shortened to addition mod 2, and it turns out to be most useful. [Ex 1]

Many practical codes are binary, but some have alphabets with more than two symbols. If, for example, the alphabet has five symbols, then in order to obtain a useful structure for such a code it is usual to take the alphabet to be $\{0, 1, 2, 3, 4\}$ and the arithmetic relevant to the code would be modulo 5.

It is time to be less vague. We can add, subtract or multiply any two integers and do these processes modulo any positive integer. To define what we have previously called the 'remainder' note that if a is any integer and b is any non-zero integer there are many ways of expressing the result of dividing a by b. For example,

$$\begin{array}{llrlrl} 10 \div 3 & \text{is} & 3 & \text{with a remainder of} & 1 & (10 = 3 \times 3 \quad +1) \\ & \text{or} & 4 & \text{with a remainder of} & -2 & (10 = 4 \times 3 \quad -2) \\ & \text{or} & 1 & \text{with a remainder of} & 7 & (10 = 1 \times 3 \quad +7) \\ & \text{or} & -3 & \text{with a remainder of} & 19 & (10 = -3 \times 3 \quad +19) \end{array}$$

But if, as is usual, we specify that the remainder must be the smallest possible non-negative value, this fixes the remainder uniquely. [Ex 2]

3.2 Congruence and related ideas

Definition 3.1 If a and b are integers and m is any positive integer, a is said to be *congruent* to b modulo m if a and b differ by a multiple of m.

The notation for this is $a \equiv b \bmod m$, and by *multiples* of m we mean the product of m with *any* integer, so 5, 25, 40, 0, -10, ... are all multiples of 5.

An equivalent way of expressing $a \equiv b \bmod m$ is to say a and b leave the same remainder on division by m. The fact that any integer must be congruent to 0, 1, 2, ..., or $m-1 \bmod m$ is often called *the division algorithm* and its use often shortens arguments considerably, as for example in Exercises 6, 7, 9 and 10 below.

It is important to note that fractions have no place in the new type of arithmetic we are about to investigate, so a, b, m or any other letter will, until the end of section 3.6, always stand for integers.

The sign for congrence, $\equiv$, is close to the sign $=$, for equality, and this is no accident for the two relations share many properties. Just how close they are is illustrated by the next theorem which gives a list of properties of $\equiv$.

Theorem 3.1 If m is any positive integer and a, b, c, d are any integers, then:

(i) $a \equiv a \bmod m$;
(ii) if $a \equiv b \bmod m$ then $b \equiv a \bmod m$;
(iii) If $a \equiv b \bmod m$ and $b \equiv c \bmod m$ then $a \equiv c \bmod m$;
(iv) If $a \equiv b \bmod m$ and $c \equiv d \bmod m$ then $a + c \equiv b + d \bmod m$ and $ac \equiv bd \bmod m$.

Proof. exercise. □

[Ex 3–7]

One method of solving a pair of simultaneous linear equations such as

$$\left.\begin{array}{rcrcl} 3x & + & 2y & = & 6 \\ 4x & - & 2y & = & 7 \end{array}\right\}$$

is to 'add the equations' so that the y-terms vanish and we are left with $7x = 13$. The first part of the property (iv) above tells us that we can do the same with congruences.

If we take the special case of (iv) in which $c = d$ we obtain

$$\begin{array}{rlclcll} & a \equiv b \bmod m & \Rightarrow & a + c & \equiv & b + c & \bmod m \\ \text{and} & a \equiv b \bmod m & \Rightarrow & ac & \equiv & bc & \bmod m \end{array}$$

The first of these implications is that it is legitimate to 'add the same thing to both sides of a congruence', and it works in reverse too: that is

$$a + c \equiv b + c \bmod m \quad \Rightarrow \quad a \equiv b \bmod m,$$

since this is just 'adding $-c$ to both sides'.

Unfortunately 'multiplying both sides by the same thing' doesn't always work in reverse. For example

$3 \times 5 \equiv 3 \times 13 \bmod 12$ is true, but 'cancelling the 3' would
give $5 \equiv 13 \bmod 12$ which is false.

This is a pity because cancelling a common factor (other than 0) is a perfectly legitimate thing to do when manipulating *equations*. We shall have reason to solve the occasional congruence in connection with codes, so we now start working towards a theorem which tells us that we can do cancellation provided we make a suitable adjustment to the modulus. First we need a definition and some important notation.

The symbol $a|b$ which number theorists usually express as 'a divides b' means a is a factor (or divisor) of b, or equivalently, b is a multiple of a. Provided r and s are not both zero the symbol $\gcd(r, s)$ denotes the greatest common divisor (also called the highest common factor) of r and s. Just as $\neq$ negates the equality relation, we can do the same with $|$ and $\equiv$. Some examples are given below and you should satisfy yourself that they are all true.

$12 \mid 360$;	$36 \nmid 12$;	$3 \not\equiv 10 \bmod 2$; $0 \mid 0$;	$0 \nmid 2$;
$\gcd(7, 21) = 7$;	$\gcd(7, 22) = 1$;	$\gcd(18, -42) = 6$;	
$\gcd(-14, -22) = 2$;	$50 \equiv 194 \bmod 12$;	$10 \equiv -23 \bmod 11$;	
$10 \not\equiv 23 \bmod 11$;			

and for all x: $3 \nmid 3x + 2$; $x \mid x$; $x \mid 0$; $1 \mid x$; $x \mid x$;
$\gcd(0, x) = x$ (provided $x \neq 0$).

[Ex 8–12]

Theorem 3.2 If $a = bq + r$ and $b \neq 0$, then $\gcd(a, b) = \gcd(b, r)$.

Proof. Let $D_{x,y}$ denote the set of all common divisors of x and y. We show that $D_{a,b} = D_{b,r}$.

First,

$$\begin{aligned} d \in D_{a,b} &\Rightarrow (a = kd \text{ and } b = ld) \Rightarrow r = a - bq = d(k - lq) \\ &\Rightarrow d \mid r \\ &\Rightarrow d \in D_{b,r} \end{aligned}$$

$$\text{so } D_{a,b} \subseteq D_{b,r}. \qquad (1)$$

Secondly,

$$\begin{aligned} e \in D_{b,r} &\Rightarrow (b = ue \text{ and } r = ve) \Rightarrow a = bq + r = e(uq + v) \\ &\Rightarrow e \mid a \\ &\Rightarrow c \in D_{a,b} \end{aligned}$$

$$\text{so } D_{b,r} \subseteq D_{a,b}. \qquad (2)$$

So combining (1) and (2) we have $D_{a,b} = D_{b,r}$. Now if two finite sets of integers are equal their greatest members must be the same!

That is

$$\gcd(a, b) = \gcd(b, r).$$

□

This theorem has the following important and famous corollory.

Euclid's algorithm

This is best explained by a typical example. We begin with any two positive integers, say 3840 and 1404, and divide the larger by the smaller to obtain the quotient 2 and remainder 1032. Then divide 1404 by 1032 to get its quotient, 1 and remainder 372. Then divide 1032 by 372, and so on. The successive steps are shown below, where the remainders are chosen in accordance with the specification at the end of section 3.1, and to the right of each division we have written the result of applying Theorem 3.2.

$3840 \div 1404.$	$3840 = 2 \times 1404 + 1032 \ldots (1),$	$\gcd(3840, 1404) = \gcd(1404, 1032)$
$1404 \div 1032.$	$1404 = 1 \times 1032 + 372 \ldots (2),$	$= \gcd(1032, 372)$
$1032 \div 372.$	$1032 = 2 \times 372 + 288 \ldots (3),$	$= \gcd(372, 288)$
$372 \div 288.$	$372 = 1 \times 288 + 84 \ldots (4),$	$= \gcd(288, 84)$
$288 \div 84.$	$288 = 3 \times 84 + 36 \ldots (5),$	$= \gcd(84, 36)$
$84 \div 36.$	$84 = 2 \times 36 + \boxed{12} \ldots (6),$	$= \gcd(36, 12)$
$36 \div 12.$	$36 = 3 \times 12 + 0 \ldots (7),$	$= \gcd(12, 0)$
		$= \boxed{12}$

The process is terminated once a zero remainder is reached, in this case after seven divisions. Clearly there is nothing special about the numbers 3840 and 1404, so we could start with any pair of integers and the result would be that their gcd is the last non-zero remainder. If you are a good sceptic, and all mathematicians should be, you will be asking 'but what if we never reach a zero remainder?' Well, suppose the sequence of remainders is $r_1, r_2, r_3, \ldots$. To obtain r_2 we do a division by r_1 so $r_2 < r_1$. To obtain r_3 we do a division by r_2 so $r_3 < r_2$, ... and so on. In other words the sequence of remainders is strictly decreasing, and any strictly decreasing sequence of integers bigger than or equal to zero must clearly reach zero eventually.

What is striking about the algorithm is that 'eventually' is very soon. For example, we needed only seven steps starting from 3840. A rough explanation is that when dividing by x we would expect 'on average' that the remainder lies midway between 0 and x, so the average behaviour of the algorithm would be to halve the remainder at each step. So starting from 3840 the expected number of steps is about eleven or twelve. [Ex 13, 14]

Much more of a surprise is that the algorithm finds the highest common factor of a pair of integers without factorizing either of them!

It can also be used to give a definitive answer to the question raised in Exercise 14. The next theorem is the major step towards this.

Theorem 3.3 If $d = \gcd(a, b)$ then d can be expressed as an integer linear combination of a and b. That is, $d = ax + by$ for some integers x and y.

Proof. Again an example will suffice. We start with another instance of Euclid's algorithm:

$$\begin{array}{rclcrcr l}
693 & = & 1 & \times & 392 & + & 301 & \ldots (1) \\
392 & = & 1 & \times & 301 & + & 91 & \ldots (2) \\
301 & = & 3 & \times & 91 & + & 28 & \ldots (3) \\
91 & = & 3 & \times & 28 & + & \boxed{7} & \ldots (4) \\
28 & = & 4 & \times & 7 & + & 0 &
\end{array}$$

So $\gcd(693, 392) = 7$.

Now starting from equation (4) and working backwards we have

$$\begin{aligned} \gcd(693, 392) = 7 &= 91 - (3 \times 28) & \\ &= 91 - 3(301 - (3 \times 91)) & \text{from (3)} \\ &= (10 \times 91) - (3 \times 301) & \text{tidying the line above} \\ &= 10(392 - (1 \times 301)) - (3 \times 301) & \text{from (2)} \\ &= (10 \times 392) - (13 \times 301) & \text{tidying} \\ &= (10 \times 392) - 13(693 - (1 \times 392)) & \text{from (1)} \\ &= (23 \times 392) - (13 \times 693) & \text{tidying} \end{aligned}$$

So by running Euclid's algorithm backwards we have expressed the gcd as an integer linear combination of the original two numbers. □

[Ex 15]

Now look again at the example in Exercise 14.

$$T_{24,42} = \{24x + 42y \ : \ x \in Z, y \in Z\}$$

We have now from Theorem 3.3 that 6, the gcd of 24 and 42, must be a member of $T_{24,42}$. And from the solution to Exercise 14 $T_{24,42}$ must contain all the multiples of 6. Furthermore, it is easy to show that it contains nothing else because any member, $24x + 42y$, can be written as $6(4x + 7y)$ which is clearly a multiple of 6. So we have proved our next result:

Theorem 3.4 For any integers a, b, not both zero, the set of all integer linear combinations of a and b is the set of all multiples of their greatest common divisor.

Definition 3.2 A pair of integers is called a *relatively prime* pair or a *coprime* pair if they have no positive common divisor except 1. (So (a, b) is coprime means $\gcd(a, b) = 1$).

[Ex 16]

Theorem 3.5 (Euclid's lemma) If $a|bc$ and (a, b) is a coprime pair, then $a|c$.

Proof.

$$\begin{aligned} a|bc &\Rightarrow bc = am & \text{for some } m \in Z \quad \ldots (1) \\ (a,b) \text{ coprime} &\Rightarrow ax + by = 1 & \text{for some } x, y \in Z \\ &\Rightarrow acx + bcy = c & \\ &\Rightarrow acx + amy = c & \text{from (1)} \\ &\Rightarrow a(cx + my) = c & \\ &\Rightarrow a|c & \end{aligned}$$

□

[Ex 17–19]

Now we can obtain the result promised some pages back.

Theorem 3.6 If $ax \equiv ay \bmod m$, then $x \equiv y \bmod \frac{m}{d}$ where $d = \gcd(a, m)$.

Proof. The result clearly holds when $a = 0$, so now deal with the case $a \neq 0$.

$$ax \equiv ay \bmod m \quad \Rightarrow \quad a(x-y) = km \qquad \text{for some } k \in Z \ldots (1)$$
$$\gcd(a,m) = d \quad \Rightarrow \quad a = ud, m = vd, \qquad \ldots (2)$$
$$\text{and } \gcd(u,v) = 1 \qquad \ldots (3)$$

Hence, from (1) and (2) $ud(x-y) = kvd.$

$d \neq 0$ so $u(x-y) = kv.$

i.e. $x - y = \frac{kv}{u}$ ... (4)

Now the left hand side of (4) is clearly an integer, so $u|kv$, and by applying Euclid's lemma with (3) we deduce $u|k$.

Hence $\frac{k}{u}$ is an integer so (4) says that $x - y$ is a multiple of v.

That is $x \equiv y \bmod \frac{m}{d}$ as required. □

In particular, this theorem says that if the modulus and the common factor we wish to cancel are coprime, then the modulus is still m after cancelling.

3.3 Solving linear congruences

Consider first the familiar quadratic equation $ax^2 + bx + c = 0$ where a, b and c are given and x is an 'unknown'. For this equation and most others the following questions are natural ones to ask.

1. Are there any solutions?
2. If so, how many?
3. Is there a method of finding one?
4. What about finding them all?

You will be familiar with the fact that the answers depend on whether a, b, c and x take real or complex values, or range over some other set. We can ask the same questions concerning *congruences*, in particular *linear congruences in a single variable*. That is, those of the form

$$ax \equiv c \quad \text{mod} m \qquad (*)$$

It is now assumed, of course, that a and c are given integers, m is a given positive integer and x is to be sought in Z.

To make a start, notice that (*) is equivalent to an *equation*, since its meaning is that $ax - c$ is a multiple of m. That is,

$$ax - my = c \text{ holds for some integers } x, y \qquad (**)$$

The original one variable problem seems to have been magically transformed into one involving two variables, but this is illusory since each solution x of (*) corresponds to just one solution (x, y) of (**), but we happen to be only interested in the x values.

If x, y, a, m, c are real numbers then all questions concerning (**) are answered at a stroke: there are infinitely many solutions and they consist of all the points (x, y) on the straight line graph of $ax - my = c$!

By restricting the numbers to be integers the problem is much more interesting as we are now asking whether this straight line passes through any points whose co-ordinates are both integers. It may or may not, depending on the values of a, c and m. The problem in this form was studied by the Greeks long before Gauss invented the notion of congruence. Indeed (**) is an example of a *Diophantine equation*, named in honour of Diophantus of Alexandria who worked on number theory in the third century AD.

This device of transforming (*) to (**) will answer questions 1 and 3: $ax - my$, and hence c, is a member of the set $T_{a,m}$ introduced in Exercise 14, so by Theorem 3.4 c has to be a multiple of $\gcd(a, m)$. This in turn means that question 1 has the following answer.

Theorem 3.7 The linear congruence $ax \equiv c \bmod m$ has a solution if and only if $\gcd(a, m)|c$.

As for question 3 we just use Euclid's algorithm to find a solution (x_0, y_0) of $ax - my = d$ where $d = \gcd(a, m)$, as in the solution of Exercise 15.

$$\text{so } ax_0 - my_0 = d \tag{3.1}$$

Now $d|c$ so $c = kd$ for some $k \in Z$, so from (1) we get

$$a(kx_0) - m(ky_0) = kd = c$$

so (kx_0, ky_0) is a solution of (**).

Now for question 2. In one sense the answer is trivial: there must be infinitely many because if x satisfies $ax \equiv c \bmod m$ then $x + lm$ also satisfies it for all integers l, as you showed in Exercise 18. To make the question more interesting (and more in tune with the spirit of number theory) we reformulate it as 'how many solutions are there if we do not regard solutions differing by a multiple of m as different?' In other words, how many solutions are there which are *distinct mod m*? Clearly at most m, but how many exactly?

Theorem 3.8 If the congruence $ax \equiv c \bmod m$ has a solution, it has precisely d solutions which are distinct mod m, where $d = \gcd(a, m)$.

Proof. Let x_0 be any particular solution, so that $ax_0 \equiv c \bmod m$.

Then $x_0 + \frac{km}{d}$ is also a solution since

$$\begin{aligned} a\left(x_0 + \frac{km}{d}\right) = ax_0 + \frac{kam}{d} &\equiv c + \frac{kam}{d} && \text{because } ax_0 \equiv c \bmod m \\ &\equiv c \bmod m && \text{since } \frac{a}{d} \text{ is an integer} \end{aligned}$$

Now consider the d consecutive solutions, making up the set

$$S = \left\{ x_0, x_0 + \frac{m}{d}, x_0 + \frac{2m}{d}, \ldots, x_0 + \frac{(d-1)m}{d} \right\}$$

The smallest and largest differ by only $\frac{(d-1)m}{d}$ which is clearly less than m, so each pair of distinct members of S will differ by less then m, so cannot be congruent mod m.

We get nothing new (mod m) by extending S onwards from $x_0 + \frac{(d-1)m}{d}$ nor backwards from x_0 because the subsequent and previous blocks of d consecutive solutions just repeat (mod m) the numbers in S. The situation is illustrated in figure 3.1.

We have now shown that there are just d distinct solutions (mod m) of the particular form $x_0 + \frac{km}{d}$. It remains to show that there are no others, but this is easy because if x' is *any* solution we have

$$ax_0 \equiv c \bmod m \text{ and } ax' \equiv c \bmod m$$

so $\quad a(x' - x_0) \equiv 0 \bmod m.$

Hence $\quad x' - x_0 \equiv 0 \bmod \frac{m}{d}$ by Theorem 3.5,

so $\quad x' = x_0 + \frac{km}{d}$

□

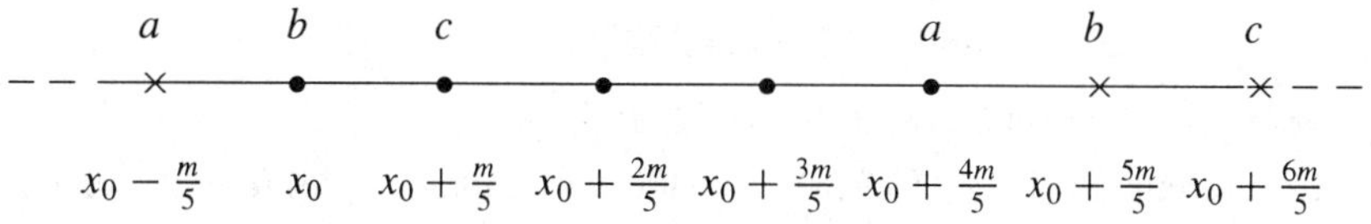

The pairs marked (a, a), (b, b), (c, c) are congruent mod m.

Figure 3.1 *The 5 distinct solutions of* $ax \equiv c \bmod m$. *(for* $\gcd(a, m) = 5, 5|c$*)*

[Ex 20]

Finally, we deal with question 4. If you refer to the solution of the final bit of Exercise 15 and combine it with our last theorem, you will have the answer. Suppose our congruence is $25x \equiv 15 \bmod 35$. This has a solution since $\gcd(25, 35) = 5$ and $5|15$. The congruence is equivalent to the Diophantine equation $25x + 35y = 15$, for which you found the particular solution $x_0 = 9$ in Exercise 15. It now follows from the proof of Theorem 3.8 that a complete set of solutions is $\{9,16,23,30,37\}$.

When the numbers involved are fairly small, as in this example, going through the whole business of Euclid's algorithm and then reversing it is rather like using a sledge-hammer to crack a nut. Instead, a judicious use of the various properties of congruences now at our disposal is often effective.

The following examples illustrate this. In all cases the congruence on each line is equivalent to the one on the line above in the sense of having the same infinite set of integers as solutions.

1. $$\begin{aligned} 48x &\equiv 18 \bmod 30 \text{ } (\gcd(48,30) = 6 \text{ and } 6|18, \text{ so 6 solutions}) \\ 8x &\equiv 3 \bmod 5 \text{ (Theorem 3.5)} \\ 3x &\equiv 3 \bmod 5 \text{ } (8 \equiv 3 \bmod 5 \text{ so } 8x \equiv 3x \bmod 5) \\ x &\equiv 1 \bmod 5 \text{ (Theorem 3.5)} \end{aligned}$$

   Note that this final congruence has only one solution modulo 5, but the original congruence to be solved was one with six distinct solutions modulo 30. These can be obtained simply by adding multiples of 5 to 1, so we obtain $\{1, 6, 11, 16, 21, 26\}$ as a complete set of solutions mod 30.

2. (same congruence, different strategy)

$$\begin{aligned} 48x &\equiv 18 \bmod 30 \\ 48x &\equiv 48 \bmod 30 \text{ (since } 18 \equiv 48 \bmod 30) \\ x &\equiv 1 \bmod 5 \text{ (Theorem 3.5)} \end{aligned}$$

3. Use negative integers if this helps to reduce the number of steps.

$$\begin{aligned} 14x &\equiv 12 \bmod 16 \text{ } (\gcd(14,16) = 2 \text{ and } 2|12, \text{ so 2 solutions}) \\ -2x &\equiv -4 \bmod 16 \text{ } (-2 \equiv 14, -4 \equiv 12 \bmod 16) \\ x &\equiv 2 \bmod 8 \text{ (Theorem 3.5).} \end{aligned}$$

   So a complete set of solutions is $\{2, 10\}$.

4. A different sequence of steps can lead to an apparently different solution:

$$\begin{aligned} 14x &\equiv 12 \bmod 16 \\ 7x &\equiv 6 \bmod 8 \\ -x &\equiv 6 \bmod 8 \\ x &\equiv -6 \bmod 8 \end{aligned}$$

   So a complete set of solutions is $\{-6, 2\}$.

   But this is of course equivalent to the previous set as $-6 \equiv 10 \bmod 16$. [Ex 21]

3.4 A bit of arithmetic folklore

Here is a well-known trick which is easily explained using congruences. Any integer can be tested for divisibility by 9 by adding its digits. The integer is divisible by 9 if and only if its digit sum is. In fact we can say more: the remainder when the number is divided by 9 is the same as the remainder when the digit sum is divided by 9.

To see that this is valid it is only necessary to notice that $10 \equiv 1 \bmod 9$, and that our ordinary notation for numbers just amounts to writing them as sums of multiples of powers of ten. How these facts are used is shown in the example below:

$$\begin{aligned} 4730289 &= (4\times10^6)+(7\times10^5)+(3\times10^4)+(0\times10^3)+(2\times10^2)+(8\times10^1)+(9\times10^0)\\ &\equiv (4\times1^6)+(7\times1^5)+(3\times1^4)+(0\times1^3)+(2\times1^2)+(8\times1^1)+(9\times1^0) \mod 9\\ &= 4+7+3+0+2+8+9\\ &= \text{digit sum of original number}\\ &= 33\\ &\equiv 6 \mod 9\end{aligned}$$

So 4730289 leaves a remainder of 6 on division by 9.

If you are really lazy and want to reduce the numbers involved as much as possible you could process the digit sum still further and continue as follows:

$$\begin{aligned} & 4 + 7 + \underbrace{3 + 0 + 2} + 8 + 9 & \\ \equiv\ & \underbrace{4 - 2 \quad + 5} \quad - 1 + 0 & \pmod 9\\ \equiv\ & \underbrace{-2 \qquad - 1} & \pmod 9\\ \equiv\ & -3 & \pmod 9\\ \equiv\ & 6 & \pmod 9\end{aligned}$$

[Ex 22–26]

3.5 The special rôle of primes

One almost obvious fact which makes prime numbers special is that every integer bigger than 1 can be expressed as a product of primes, for example $10164 = 2^2 \times 3^1 \times 7^1 \times 11^2$ (see Theorem 3.9 below). So primes are the bricks from which all positive integers (except 1) are built. We shall see later that codes in which the alphabet size is prime have particularly useful properties.

Definition 3.3 A prime number is a positive integer which has exactly two positive divisors.

Thus, 2 and 17 are primes because their only positive divisors are 1, 2 and 1, 17 respectively; 1 is not since 1 is its only positive divisor; 15 is not since it has too many, 1,3,5 and 15.

The positive integers are conveniently split into three categories:

1 – which is special and goes in a class of its own;
the primes – which we have just defined;
the composite numbers – which are all the rest.

Notice that composite numbers can be characterized as those positive numbers which can be expressed as the product of two positive factors neither of which is 1.

Theorem 3.9 Every composite number is a product of primes.

Proof. (by contradiction)

If the theorem is false then there is at least one composite number not expressible as a product of primes. Let n be the smallest of these.

Then $n = ab$ with $a \neq 1, b \neq 1$.

Hence $a < n$ and $b < n$, so a and b are products of primes.

Hence ab is a product of primes – a clear contradiction. □

The next theorem guarantees an unlimited supply of primes, and of the many proofs, the easiest is similar to Euclid's original one, made even easier by the language of congruence.

Theorem 3.10 There are infinitely many primes.

Proof. Let $F = \{p_1, p_2, \ldots p_n\}$ be any finite non-empty set of primes.

Consider the number $N = p_1 p_2 \ldots p_n + 1$.

N is clearly bigger than each p_i so $N \notin F$.

So if N is prime we have found a prime outside F. If not, it must be composite, so by Theorem 3.9 it has a prime factor p. Now $N \equiv 1 \bmod p_i$ for each p_i in F, but $N \equiv 0 \bmod p$, so p is a prime outside F.

So in either case we have found a prime outside F. But F was *any finite* set of primes, so we have proved that no finite set of primes can contain all of them.

Hence the set of primes must be infinite. □

For comments on why this result is not as obvious as it seems see [4]. The same reference contains a discussion of how the subtlety of the next result is so well hidden that it took the genius of Gauss to realize even that a proof was necessary. So much depends upon it that it is often dignified by the title of *Fundamental Theorem of Arithmetic.*

Theorem 3.11 For each integer $n > 1$ there is only one collection of primes whose product is n.

Preliminary comments:

1. If n is prime then of course the collection consists of just n itself.
2. For our example at the start of this section the theorem declares that to make 10164 as a product of primes we must have a pair of 2s, a pair of 11s, a 3 and a 7. We cannot vary the primes involved nor the number of times each one occurs. Writing them down in a different order does not count as a different factorization, so $2^2 \times 3 \times 7 \times 11^2, 11^2 \times 3 \times 2^2 \times 7$ and $2 \times 3 \times 7 \times 11 \times 2 \times 11$ are all regarded as the same factorization.
3. Euclid's lemma (Theorem 3.5), in particular Exercise 17(c), is what makes the proof relatively easy.
4. The main idea of the proof is to take any two collections of primes whose product is n and show they must be the same.

Proof. Suppose $n = p_1p_2 \ldots p_k = q_1q_2 \ldots q_l$ (1)
where the ps and qs are primes written in non-decreasing order.
That is $p_1 \leq p_2 \leq \ldots \leq p_k$ and $q_1 \leq q_2 \leq \ldots < q_l$.

Now $p_1|n$ so by Exercise 17(c) $p_1 = q_i$ for some i (2)
Similarly, $q_1|n$ so $q_1 = p_j$ for some j (3)

But $q_i \geq q_1$ and $p_j \geq p_1$, so combining these with (2) and (3) we get $p_1 \geq q_1$ and $q_1 \geq p$,

So $p_1 = q_1$.

So cancelling p_1 and q_1 from (1) we get

$$p_2p_3 \ldots p_k = q_2q_3 \ldots q_l.$$

Applying the same argument again we deduce that $p_2 = q_2$, so cancelling,
$p_3 \ldots p_k = q_3 \ldots q_l$, ... and so on.

We continue to cancel p_i with q_i until we run out of p's or q's or both, and end up with:

$$\begin{array}{rll} 1 & = q_{k+1}q_{k+2} \cdots q_l & \text{if } l > k \\ p_{l+1}p_{l+2} \cdots p_k & = 1 & \text{if } k > l, \quad \text{or} \\ 1 & = 1 & \text{if } k = l. \end{array}$$

Clearly the first two cases are impossible, which leaves the final case in which all the ps and qs have cancelled in pairs. That is, the two collections of primes were in fact the same. □

[Ex 27–29]

One of the many results which Fermat announced in his letters to mathematical correspondents in Europe has become known as Fermat's Little Theorem. It appeared in 1640 although the first published proofs seem to be those of Leibniz and Euler. Today the language of congruence makes its statement succinct and its proof easy. We also make use of Exercise 20.

Theorem 3.12 If p is prime and a is not a multiple of p then $a^{p-1} \equiv 1 \bmod p$.

Proof. The set $\{0, 1, 2, \ldots, p-1\}$ is clearly a complete residue set mod p. From the conditions of the theorem $\gcd(a, p) = 1$, so by Exercise 20, $\{0, a, 2a, \ldots (p-1)a\}$ is also a complete residue set mod p.

Hence only the first member is congruent to 0 mod p, so the rest are congruent to $1, 2, \ldots, p-1$ (though not necessarily in that order).

So $a \times 2a \times 3a \times \ldots \times (p-1)a \equiv 1 \times 2 \times 3 \times \ldots \times (p-1) \bmod p$.

That is $a^{p-1}(p-1)! \equiv (p-1)! \bmod p$ (1)

Now p is prime, so $1, 2, 3, \ldots, p-1$ are all coprime with p, so $\gcd(p, (p-1)!) = 1$, and cancellation of the $(p-1)!$ from (1) is valid.

Hence $a^{p-1} \equiv 1 \bmod p$ as required. □

3.6 A recreational interlude

The previous proof is a purely number-theoretic one. It is also possible to give a more 'visual' combinatorial proof as follows.

Suppose you have an unlimited supply of beads of a different colours and you are making decorations consisting of vertically hanging chains of p beads each. There are a^p distinct chains which can be made (just the number of words of length p with an alphabet of size a again), and a of these will contain beads all of the same colour. So $a^p - a$ is the total number of possible decorations containing at least two colours each. Clearly the intention is to count two decorations as 'the same' if and only if they contain the same collection of colours in the same order.

Now consider any one of these and transform it by taking the top bead and replacing it on the bottom, so that, for example,

A		B
B		C
C	becomes	D
D		E
E		A

We shall refer to each application of this process as a 'beheading'. If beheading is repeated we are clearly back to the original pattern after doing it p times. But in some cases we could achieve this in fewer than p applications. For example,

A	
B	
C	only requires three beheadings in order
A	to restore the original pattern.
B	
C	

But because we have excluded chains of a single colour we can never get back to the original with only one beheading. Now let k be the *smallest positive* number of successive beheadings needed to get back to the original. So we know $1 < k \leq p$.

Divide p by k to get

$$p = qk + r \qquad (0 \leq r < k) \tag{1}$$

Notice that k 'reverse beheadings' will also transform the original pattern into itself, where a reverse beheading is defined in the obvious way, as

$$\begin{array}{ccc} \mathrm{A} & & \mathrm{F} \\ \mathrm{B} & & \mathrm{A} \\ \mathrm{C} & \text{reverse beheading} & \mathrm{B} \\ \mathrm{D} & \longrightarrow & \mathrm{C} \\ \mathrm{E} & & \mathrm{D} \\ \mathrm{F} & & \mathrm{E} \end{array}$$

Now consider equation (1) in the form $r = p - qk$ and contemplate doing $p - qk$ successive beheadings, which it is helpful to think of as p beheadings followed by q lots of k reverse beheadings. The result is clearly to restore the original pattern, and the sequence of moves is, from (1), equivalent to just doing r beheadings!

But k was the *smallest* positive number of beheadings which would do this, and since $0 \leq r < k$ this can only mean $r = 0$, so (1) becomes $p = qk$.

Now look at the case of p being prime: k was greater than 1 so we must have $k = p$ and $q = 1$. Take any one of the chains and its first $p - 1$ beheadings, and call this collection C_1. So C_1 contains

$$\begin{array}{ccccc} b_1 & b_2 & b_3 & & b_p \\ b_2 & b_3 & b_4 & & b_1 \\ b_3 & b_4 & b_5 & \cdots\cdots & b_2 \\ \vdots & \vdots & \vdots & & \vdots \\ \vdots & \vdots & b_p & & \vdots \\ \vdots & b_p & b_1 & & \vdots \\ b_p & b_1 & b_2 & & b_{p-1} \end{array}$$

and these are all different because if not, one of them could be transformed into another by fewer than p beheadings.

If there is any other decoration *not* in C_1, select one of them and let C_2 be the collection consisting of this one and its successive $p - 1$ beheadings. These are also clearly all different and no member of C_2 can occur in C_1.

Continuing in this way until all the $a_p - a$ chains have been accounted for, suppose we get a total of n collections. Each collection contains p decorations so $a^p - a = np$.

This means $a^p \equiv a \bmod p$.

Finally, if $\gcd(a, p) = 1$ we can cancel the a from this congruence to obtain Fermat's theorem.

Notice that until this very last step no assumption had been made concerning the number a, the number of colours available, so the result $a^p \equiv a \bmod p$ is true for *any prime* p and *any* a.

The collections C_1, C_2, $\ldots$ with $a = 4, p = 3$ are shown in Figure 3.2.

○□△	○△□	○○□	□□△	△△+	++○
□△○	△□○	○□○	□△□	△+△	+○+
△○□	□○△	□○○	△□□	+△△	○++
○□+	○+□	○○△	□□+	△△○	++□
□+○	+□○	○△○	□+□	△○△	+□+
+○□	□○+	△○○	+□□	○△△	□++
○△+	○+△	○○+	□□○	△△□	++△
△+○	+△○	○+○	□○□	△□△	+△+
+○△	△○+	+○○	○□□	□△△	△++
□△+	□+△	12 collections using 2 colours			
△+□	+△□				
+□△	△□+				

8 collections
using 3 colours

Figure 3.2 *The 20 collections of decorations of length 3 with 4 colours available (indicated by 4 different shapes)* $a^p - a = 4^3 - 4 = 60$.

Fermat's theorem can sometimes help to evaluate large powers modulo some prime, as the following example illustrates.

Working modulo 37,

$$\begin{array}{rcccccl}
54^{99} & = & ((54)^{36})^2.54^{27} & \equiv & 1 \times 54^{27} & & \text{by Fermat's theorem} \\
& \equiv & (-20)^{27} & = & (-20)((-20)^2)^{13} & = & (-20)(400)^{13} \\
& \equiv & (-20)(-7)^{13} & = & (140)(-7)^{12} & \equiv & (-8)(49)^6 \\
& \equiv & (-8)(12)^6 & = & (-8)(144)^3 & \equiv & (-8)(-4)^3 \\
& = & 8 \times 64 & \equiv & 8(-10) & = & -80 \equiv -6 \\
& \equiv & 31 & & & &
\end{array}$$

[Ex 30–34]

A recreational application of Fermat's theorem is to the theory of card shuffling. The Faro shuffle is one in which a normal pack of 52 cards is split into two equal piles and the cards from the two piles are then interleaved. The process is shown in Figure 3.3 below.

You can think of the cards in their original order being labelled 1 – 52. After the split cards 1 – 26 go in the first pile and cards 27 – 52 in the second pile. The interleaved pack starts with the first card of pile 2 as its top card.

In the column headed * we have written down the 52 positions again, and then in the column headed + we have increased by 53 each of the

Original pack		Pile 1	Pile 2		New Pack	*	+
1		1	27		27	1	54
2		2	28		1	2	2
3		3	29		28	3	56
4		4	30		2	4	4
5		⋮	⋮		29	5	58
6	split pack	⋮	⋮	Interleaf	3	6	6
⋮	⟶	⋮	⋮	⟶	⋮	⋮	⋮
⋮	⋮	⋮	⋮	⋮	⋮	⋮	⋮
⋮		24	50		⋮	⋮	⋮
⋮		25	51		⋮	⋮	⋮
⋮		26	52		51	49	102
50					25	50	50
51					52	51	104
52					26	52	52

Figure 3.3 *The Faro Shuffle.*

odd numbers in * and left the even ones alone. So if i and j are two corresponding numbers in columns *, + respectively, $i \equiv j \bmod 53$.

The numbers in the 'new pack' and * columns can be interpreted as 'the card originally in position 27 gets moved to position 1, the card originally in position 1 gets moved to position 2, ... and so on'. Finally, taking columns 'new pack' and + we see that, *modulo 53*, the effect of the Faro shuffle is to double the position of each card. So the card in position x before the shuffle will be in position $2x$ modulo 53 after the shuffle. If the shuffle is done repeatedly, say n times, card x will end up in position $2^n\, x \bmod 53$.

Now for the problem: how many times does the Faro shuffle need to be repeated in order for all the cards to return to their original positions? So we require the smallest positive n for which

$$x \equiv 2^n\, x \bmod 53 \text{ for all } x \text{ in the range } 1 \leq x \leq 52$$

This clearly holds if and only if

$$1 \equiv 2^n \bmod 53$$

Now 53 is prime and $\gcd(2, 53) = 1$ so Fermat's theorem supplies the perhaps not too surprising conclusion that 52 shuffles will do it. But we still do not know whether we can do it with fewer shuffles. Here the division algorithm will help to drastically reduce the number of possibilities to try.

Suppose $k(1 \leq k \leq 52)$ is the smallest number of shuffles to restore the original pack. Divide 52 by k to get $52 = qk + r, 0 \leq r < k$.

Then $2^{52} = 2^{qk+r} \equiv 1 \bmod 53$.

That is: $(2^k)^q.2^r \equiv 1 \bmod 53$, and since by hypothesis $2^k \equiv 1 \bmod 53$, we have

$$2^r \equiv 1 \bmod 53$$

and since $0 \leq r < k$ and k is the least positive solution this can only mean $r = 0$.

Hence $52 = qk$ so the only candidates for k are 1, 2, 4, 13, 26 and 52. Clearly k =1, 2 or 4 fails to satisfy $2^k \equiv 1 \bmod 53$. We leave you to check that $2^{26} \not\equiv 1 \bmod 53$, and hence $2^{13} \not\equiv 1 \bmod 53$ because $2^{26} = (2^{13})^2$.

So 52 shuffles restores the pack order, and no smaller number will do it.

The solution of Exercise 31 may help with the next exercise, and for Exercise 36 it is easier to number the original pack as 0, 1, 2, 3, ... rather than 1, 2, 3, 4, ... [Ex 35, 36]

3.7 Z_p and reciprocals

To conclude this taste of number theory we change our point of view slightly, to highlight a fact which will be essential to our subsequent discussion of linear codes. Suppose a, b, c are integers which satisfy $a + b \equiv c \bmod m$ and a', b', c' are integers which are congruent mod m to a, b, c respectively, then $a' + b' \equiv c' \bmod m$. Similarly, if $ab \equiv c \bmod m$, then $a'b' \equiv c' \bmod m$. [Ex 37]

These results enable us to do consistent arithmetic modulo m, not with individual integers, but with classes of integers. If we let $[x]$ stand for the class of all integers congruent to x mod m, then Z is partitioned into the m non-overlapping classes $[0], [1], [2], \ldots, [m-1]$. Using the result above we can deduce, from for example the congruence $5 + 9 \equiv 4 \bmod 10$, that

any integer in [5]+ any integer in [9] $\equiv$ every integer in [4] mod 10. (1)

Similarly for multiplication;

from $7 \times 9 \equiv 3 \bmod 10$, we get:

any integer in [7]$\times$ any integer in [9] $\equiv$ every integer in [3] mod 10. (2)

The set $\{[0], [1], [2], \ldots, [m-1]\}$ is denoted by Z_m, and we have just invented 'arithmetic' in Z_m. (1) and (2) above are usually written as

$$\begin{aligned} [5] + [9] &= [4] \text{ in } Z_{10} \\ \text{and } [7] \times [9] &= [3] \text{ in } Z_{10}, \end{aligned}$$

and we often go further and write Z_m as just $\{0, 1, 2, \ldots, m-1\}$, and the foregoing relations as

$$\left.\begin{array}{rcl} 5+9 & = & 4 \\ \text{and } 7\times 9 & = & 3 \end{array}\right\} \text{ in } Z_{10}$$

and it is the result of Exercise 37 which makes this abuse of notation legitimate.

Below is a list of those rules of arithmetic with ordinary numbers which also hold in Z_m arithmetic. Most of them are obvious, and any that you are doubtful about can be checked by re-interpreting them in terms of congruences. We should preface the list with the observation that Z and Z_m are *closed* under the $+, -, \times$ operations. This means that if you add, subtract or multiply two integers the result is always another integer, and if you add, subtract or multiply two members of Z_m the result is another member of Z_m. (Without this the items in the list below would be meaningless!)

For all a, b, c in Z_m (for any fixed integer $m \geq 1$)

$$\begin{array}{lrcl}
1 & a+b & = & b+a \\
2 & (a+b)+c & = & a+(b+c) \\
3 & 0+a & = & a \\
4 & a-a & = & 0 \\
5 & ab & = & ba \\
6 & (ab)c & = & a(bc) \\
7 & 1a & = & a \\
8 & 0a & = & 0 \\
9 & a(b+c) & = & ab+ac
\end{array}$$

One of the properties of the integers which is lost in Z_m is the fact that if $ab = 0$, then at least one of a and b must be zero. To see that this fails in Z_m consider the example of Z_6 in which $2 \times 3 = 0$ but $2 \neq 0$ and $3 \neq 0$. However, if m is restricted to being *prime* this property is restored.

Theorem 3.13 If p is prime and $a, b \in Z_p$ and $ab = 0$, then a or b (or both) are zero.

Proof. The proof works by showing that if a or b is non-zero and $ab = 0$, then the other must be zero.

So suppose $ab = 0$ and $a \neq 0$.

Interpreting this as a congruence, it means that $ab \equiv 0 \bmod p$, and a is not a multiple of p.

Hence $\gcd(a, p) = 1$, so we can use Theorem 3.5 to cancel a from the congruence and obtain

$$\begin{array}{rcl} b & \equiv & 0 \bmod p \\ \text{That is } b & = & 0 \text{ in } Z_p \end{array}$$

□

Related to this is a property of Z_p (again with p prime) which the integers do not possess. The only integers which have integer reciprocals are 1 and -1. In Z_p *all* the non-zero members have reciprocals.

Theorem 3.14 If p is prime and x is a non-zero member of Z_p, then Z_p has a unique member y such that $xy = 1$.

Proof. Reinterpreting as a congruence, we are aiming to show that $xy \equiv 1 \bmod p$ has a solution for y if $x \not\equiv 0 \bmod p$. Now $\gcd(x, p) = 1$ and $1|1$, so Theorems 3.7 and 3.8 guarantee that there is a solution and that it is unique. □

[Ex 38]

The reciprocal of a in Z_p is often written as a^{-1}, and in Z_p we can do something analogous to division by interpreting $b \div a$ as ba^{-1}.

Primes are going to be vitally important in nearly all our subsequent work on codes.

3.8 Further reading

There are many books on number theory and this chapter will have given you the background to study most of those which cover a typical undergraduate course in the subject. [5] is a good general pure number theory text of this type, which also contains sketches of how its history has been influenced by many colourful characters. There is ample scope for anyone with an interest in computing to do some significant investigations in number theory and [6] is written with this sort of reader in mind. [7] also contains all the basic theory but has material on the more recent applications to computer science and to coding's sister subject, cryptography. On specific topics of this chapter, [4] considers the question of why the Fundamental Theorem of Arithmetic, Euclid's algorithm, the infinitude of the primes, etc, are surprising and/or significant. For readers interested in pursuing recreational number theory [8], [9] and [10] are interesting articles. There is even a *Journal of Recreational Mathematics*, many of whose topics are based on elementary number-theoretic ideas.

3.9 Exercises for Chapter 3

1. In section 3.1 we claimed that the 'sum' of any two codewords of the Hamming (7,4) code is another codeword of the same code. Can you think of a way of verifying this without summing all possible pairs of the sixteen codewords?
2. Suppose $a = qb + r$ is a result of dividing a by b. Show that the requirement that the remainder is ≥ 0 and $< b$ makes the values of q and r unique.

3. Prove properties i – iv listed in Theorem 3.1.
4. Show from Theorem 3.1 that $a \equiv b \bmod m \Rightarrow a^n \equiv b^n \bmod m (n \geq 0)$. How can this be used to show quickly that $41 | 2^{20} - 1$?

   Find the remainders when 2^{50} and 41^{65} are each divided by 7.

   What is the remainder when $\sum_{i=1}^{100} i^5$ is divided by 4?
5. Show that

   (a) $a \equiv b \bmod m \not\Rightarrow c^a \equiv c^b \bmod m$

   and

   (b) $a^2 \equiv b^2 \bmod m \not\Rightarrow a \equiv b \bmod m$.
6. Use the division algorithm to show that the fourth power of an integer can only be congruent to 0 or 1 mod 5.
7. For all integers n, show that $\frac{1}{6}(n+1)(2n+1)n$ is an integer.
8. Prove the following properties of $|$:

   (a) $a|b$ and $c|d \Rightarrow ac|bd$;

   (b) $a|b$ and $b|a \Leftrightarrow a = \pm b$;

   (c) $a|b$ and $b \neq 0 \Rightarrow |a| \leq |b|$;

   (d) $a|b$ and $a|c \Rightarrow a|bx + cy$ for all x, y.
9. For any integer a show that $3|a$ or $3|a+2$ or $3|a+4$.
10. If $2 \not| a$ and $3 \not| a$ prove that $24 | a^2 - 1$.
11. Prove that if c is a divisor of both a and b, then it must be a divisor of $\gcd(a, b)$.
12. Using part (d) of Exercise 8 show that for any integer n,

$$(d|2n+1 \text{ and } d|n^2+3n+1) \Rightarrow d|5n+2,$$

   and then that $(d|2n+1$ and $d|5n+2) \Rightarrow d|1$.

   Deduce that $\gcd(n^2+3n+1, 2n+1) = 1$.
13. Euclid's algorithm will take a large number of steps to deliver the final gcd if the sequence of remainders only decreases by a small number at each step. Explain in general terms why this cannot happen.
14. Let $T_{r,s}$ be the set $\{rx + sy : x \in Z, y \in Z\}$. Experiment with $T_{24,42}$ to get some idea of which integers this set contains.
15. Use Euclid's algorithm to find a solution of $1729\, x + 703\, y = 19$.

    Can you generate any more solutions?

    What about the equation $25x + 35y = 15$?
16. If $\gcd(a, b) = d$, show that $\gcd(\frac{a}{d}, \frac{b}{d}) = 1$.
17. State the special case of Theorem 3.5 which you obtain by taking a to be prime.

18. Prove that if l is any integer and x satisfies $ax \equiv c \bmod m$, then so does $x + lm$.

19. Prove the following:

   (a) if $a \equiv b \bmod n$ and $m|n$ then $a \equiv b \bmod m$;
   (b) if $a \equiv b \bmod m$ then $ca \equiv cb \bmod m$;
   (c) if $a \equiv b \bmod m$ then $\gcd(a, m) = \gcd(b, m)$.

20. A set of m integers which are distinct mod m is sometimes called a *complete residue set* mod m. Show that, mod 11, $\{0, 1, 2, 2^2, \ldots, 2^9\}$ is a complete residue set, but $\{0^2, 1^2, 2^2, \ldots, 10^2\}$ is not.

   If $\{a_1, a_2, \ldots, a_m\}$ is a complete residue set mod m, and $\gcd(k, m) = 1$, then so is $\{ka_1, ka_2, \ldots, ka_m\}$. Prove this.

21. Find, by any method, a complete set of solutions for these congruences:

(a)	$4x$	$\equiv$	5 mod 7;	(b)	$8x$	$\equiv$	12 mod 19;
(c)	$12x$	$\equiv$	3 mod 4;	(d)	$45x$	$\equiv$	75 mod 100;
(e)	$111x$	$\equiv$	112 mod 113;	(f)	$140x$	$\equiv$	133 mod 301.

22. Which properties of congruence are used in the proof of the divisibility test of section 3.4?

23.(a) Using the fact that $10 \equiv -1 \bmod 11$, devise a method of finding the remainder when any integer is divided by 11.

   (b) What is 10^3 mod 13? Use your answer to devise a divisibility test for 13.

24. n is a positive integer with an even number of digits. m is formed by moving the last digit of n to the front, so, for example, if n is 589274 then m is 458927, and if n is 7310 then m is 731.

   Prove that $11|n + m$ and $99|n^2 - m^2$.

25. Show that in the Fibonacci sequence 1, 1, 2, 3, 5, 8, 13, ... in which the first two terms are 1 and any subsequent term is the sum of its two predecessors, a term is divisible by 7 if and only if it is divisible by 21.

26. Without doing an exhaustive search, show that $x^2 + y^2 = 999$ has no integer solutions.

27. If a is composite and $a \geq 6$ prove that $a|(a - 1)!$

28. Are the following true or false? Prove the true statements and provide a counter example for the false ones (a and b are positive integers and p is prime):

   (a) if $\gcd(a, b) = p$ then $\gcd(a^2, bp) = p^2$;
   (b) if $\gcd(a, p^2) = p$ and $\gcd(b, p^2) = p^2$ then $\gcd(ab, p^4) = p^3$;
   (c) if $\gcd(a, b) = p$ then $\gcd(a^2, ab) = p^2$;
   (d) if $a^2 + b^2 = p^2$ then $\gcd(a, b) = 1$.

29.

$$\begin{array}{rcllll} 1464463 & = & 7^2\times & 11^2 & \times 13 & & \times 19 \\ 14108963 & = & & 11^2 & & \times 17 & \times 19^3 \end{array}$$

so $\gcd(1464463, 14108963) = 11^2 \times 19 = 2299$.

How does the validity of this method of finding gcds depend on the Fundamental Theorem of Arithmetic?

30. In the 'visual' proof of Fermat's theorem why is it not possible for C_1 and C_2 to have any member in common?

31. The 'visual' proof of Fermat's theorem contained the corollary that $a^p \equiv a \bmod p$ for *all* a and *all primes* p. Use this to show that $a^{25} \equiv a \bmod 195$ for all a. (Note that 195 is *not* prime.)

32. Use Fermat's theorem to help evaluate $99^{101} \bmod 31$.

33. By evaluating $2^{340} \bmod 341$ show that the converse of Fermat's theorem is false.

34. Show that $10^n \equiv 4 \bmod 6$ for all $n \geq 1$, and hence that if $m \equiv n \bmod 6$ then $10^m \equiv 10^n \bmod 7$.

From this determine the remainder when

$$10^{10} + 10^{(10^2)} + 10^{(10^3)} + \ldots + 10^{(10^{10})} \text{ is divided by 7.}$$

35. Analyse the result of repeated Faro shuffles on a pack containing two jokers (so 54 cards in all).

36. Analyse the slight variation of the Faro shuffle in which the 'new pack' starts 1, 27, 2, 28, ... – that is, we take a card from pile 1 first.

37. Prove the two claims made at the start of section 3.7.

38. Find, in Z_7, the reciprocals of 1, 2, 3, 4, 5 and 6. Use your results to solve $32x \equiv 40 \bmod 7$.

4

Block codes – some constraints and some geometry

4.1 The main problem

We saw in Chapter 1 how the 3-fold repetition code would drastically reduce the probability of messages being wrongly decoded, and discussed the price which this entailed. Then in Chapter 2 we saw how the Hamming code solved the same problem of guaranteeing to correct every instance of a single error per codeword at much less cost. The way in which it achieved its impressive performance was by ensuring that amongst the 16 7-bit codewords, no pair of them was separated by a Hamming distance less than 3. We also developed general results, Theorems 2.1 and 2.2, to connect the minimum distance of the code with its error detecting and correcting potential.

We begin this chapter by asking whether it is possible to do better than Hamming's solution. In this context a block code is characterized by its word length n, the number of codewords M, and its minimum distance d, and a standard terminology is to speak of an (n, M, d) code. If the code is not binary, we extend this to a $q - ary$ (n, M, d) code where q is the alphabet size. So the first Hamming code we discussed is an example of a (7,16,3) code.

Now n represents a 'cost' of sending each codeword, for longer codewords take longer to send than short ones; M can be regarded as a measure of the 'richness' of the language we are using; d is a measure of how accurately the code detects and corrects errors. The Hamming 16 word code is adequate for the purpose proposed in Exercise 2 of Chapter 2, but clearly not for sending messages in English (at least, not if we require a different codeword to represent each distinct letter of the alphabet). On these criteria we would be justified in calling any of the following codes 'better' than the Hamming code:

1. $n < 7,\quad M \geq 16,\quad d \geq 3;$
2. $n \leq 7,\quad M > 16,\quad d \geq 3;$
3. $n \leq 7,\quad M \geq 16,\quad d > 3.$

We shall see that there is no such code, so in this sense the Hamming (7,16,3) code is a best possible solution.

From this discussion it is apparent that in designing a good code we should aim for low n, high M and high d. Like most similar situations in life these requirements are mutually incompatible so the code we eventually settle on will have to be some sort of compromise. For example, suppose we fix n and d, then try to get as many codewords as possible. Choose a first codeword then throw in more and more, each time ensuring that the new codeword differs in at least d places from all the others. Clearly, as more new words are added it becomes harder to find the next one, and eventually there is no room for any more. (Think of packing spheres into a box.) This natural limitation on the size of M will be made quantitative in the next section.

If a code with good n, M and d values is to be *used*, it is essential to have an efficient decoding algorithm so that received messages can be quickly understood and acted upon. Without this the effort of producing a mathematically 'good' code would be wasted in practice. Nevertheless, it is still interesting to pursue a mathematical investigation of good codes, and in many cases the nice mathematics actually leads to convenient decoding procedures. The Hamming code is an example of this since the diagrammatic method of decoding, used for illustrative purposes in Chapter 2, can easily be mechanized to become an extremely rapid decoding process. This is a consequence of the code having much more mathematical structure than we have yet revealed. Enlightenment will come in Chapter 5.

4.2 Limitations on M

Let us aim to maximize M for fixed n, d and q. One method of doing this was explained by R. C. Singleton in 1964 though the result had been known for some years before this. The argument is simple and neat: imagine a $q - ary$ (n, M, d) code, C with its codewords written out as a list. Then make a new list of words, L, each of length $n - d + 1$ simply by omitting the first $d - 1$ symbols from each word of C. Let s_{ij} be the jth symbol in the ith word of C, so we have the picture shown below.

$$
\underbrace{\begin{array}{cccc}
s_{11} & s_{12} & \cdots\cdots & s_{1d-1} \\
s_{21} & s_{22} & \cdots\cdots & s_{2d-1} \\
 & & \vdots & \\
 & & \vdots & \\
 & & \vdots & \\
s_{M1} & s_{M2} & \cdots\cdots & s_{Md-1}
\end{array}
\left|
\underbrace{\begin{array}{ccc}
s_{1d} & \cdots\cdots & s_{1n} \\
s_{2d} & \cdots\cdots & s_{2n} \\
 & \vdots & \\
 & \vdots & \\
 & \vdots & \\
s_{Md} & \cdots\cdots & s_{Mn}
\end{array}}_{L}\right.}_{C}
$$

Now no two words in L can be identical as this would mean that the corresponding codewords of C could differ at most in the remaining $d-1$ places, contradicting the fact that $d(C) = d$. Hence L cannot have repeated words so M is at most the number of distinct $q-ary$ words of length $n-(d-1)$. This is q^{n-d+1} since there are q choices of symbol for each of the $n-d+1$ positions. So we have proved the result below.

Result 4.1 The Singleton bound.

$$\boxed{M \leq q^{n-d+1}}$$

Notice that this gives no information about the existence or otherwise of codes with exactly q^{n-d+1} codewords. It only says there are certainly no codes with more. In fact, for certain values of q, n and d codes for which $M = q^{n-d+1}$ do exist (the so-called maximum distance separable (MDS) codes), and their study is an interesting research topic. See Chapter 15 of [1] or Chapter 11 of [11], for example. [Ex 1]

A rather trivial upper limit to M is of course q^n since this is the total number of $q-ary$ words of length n. This upper bound is so crude as to be useless, but it does illustrate the point that different arguments can lead to valid (but different) upper bounds on M. We now use one of these different but slightly more sophisticated arguments to derive an alternative to the Singleton bound. Suppose we want a $q-ary$ code C of length n which is t-error correcting (so d must be at least $2t+1$). Geometric language will help visualization of the argument: just as a sphere of radius r in ordinary geometry is the set of all points at distance $\leq r$ from some centre point, we define $S(\boldsymbol{u}, r)$ as the analogue of this in the space of $q-ary$ words of length n as the set of all such words at Hamming distance $\leq r$ from the chosen 'centre' word $\boldsymbol{u}$:

$$S(\boldsymbol{u}, r) = \{\boldsymbol{v} : d(\boldsymbol{u}, \boldsymbol{v}) \leq r\}$$

Now the condition that C is t-error correcting is equivalent to the condition that no two spheres of radius t centred on codewords can intersect. [Ex 2]

Using this result, consider the set of all spheres of radius t centred on codewords. The total number of words is q^n, and no word belongs to more than one of the spheres, but there may, of course, be some words which don't belong to any of the spheres. So if we count the number of words in the union of all the spheres the result will be at most q^n. The number of words $\boldsymbol{v}$ at distance i from codeword $\boldsymbol{u}$ is $\binom{n}{i}(q-1)^i$ because there are $\binom{n}{i}$ choices for which i of the n places in $\boldsymbol{v}$ differ from the corresponding places in $\boldsymbol{u}$, and the symbol in each of these i places of $\boldsymbol{v}$ can be chosen in $(q-1)$ ways since the symbol can be any of the q symbols available except the one in $\boldsymbol{u}$. Hence the number of words in the sphere $S(\boldsymbol{u}, t)$ is

$$\sum_{i=0}^{t} \binom{n}{i}(q-1)^i.$$

There are M of these spheres so in the union of all of them there are $M\sum_{i=0}^{t}\binom{n}{i}(q-1)^i$ words, and this cannot exceed q^n.

This leads to the following famous result.

Result 4.2 The Hamming, or 'sphere-packing' bound.

$$\boxed{M \leq q^n \left[\sum_{i=0}^{t} \binom{n}{i}(q-1)^i\right]^{-1}}$$

For binary codes these bounds simplify to

$$M \leq 2^{n-d+1} \quad \text{and} \quad M \leq 2^n \left[\sum_{i=0}^{t} \binom{n}{i}\right]^{-1}.$$

As an example of their use, let us investigate whether there can be a binary 3-error correcting code of length 12 having at least 100 codewords. To get 3-error correction we must have $d \geq 7$, so the Singleton bound gives $M \leq 2^{12-7+1} = 64$, so the required code cannot exist.

Suppose then that we still insist on $M \geq 100$ and 3-error correction, but are prepared to compromise on n, say let n be 13. This time the Singleton bound says that with $n = 13$ and $d = 7$, M has to be ≤ 128. But note that this does not settle the question of whether such a code exists. All we can say is that the Singleton bound has not ruled it out. But we still have the Hamming bound, so does this give us stronger information? With $n = 13$ and $t = 3$ it gives

$$M \leq 2^{13}\left[1 + 13 + \binom{13}{2} + \binom{13}{3}\right]^{-1} = 21.6\ldots,$$

and M is of course an integer so $M \leq 21$. This does settle it – the required code cannot exist. [Ex 3]

It can, and often does happen, that both of our upper bounds are too weak to give conclusive answers to questions like those we have just considered. It is in the nature of *upper* bounds on M that for given values of n, d and q they can tell us that *some* values of M are impossible, and they can never tell us that a value of M *is* possible. An on-going research area is to find more powerful upper bounds for M, and we shall say a little more about it later.

You may have noticed from trying the Singleton bound to answer Exercise 3, and from its preceding paragraph, that the Hamming bound is more powerful. If this were always the case then, apart from the fact that the arithmetic involved in applying the Singleton bound is simpler, there would be no point in bothering with this bound. Let us investigate this for binary codes. Under what circumstances, if any, does the Singleton bound rule out more values of M than are ruled out by the Hamming bound? That is, we wish to solve

$$2^{n-d+1} < 2^n \left[\sum_{i=0}^{t} \binom{n}{i} \right]^{-1} .$$

Rearranging this we get

$$2^{d-1} > \sum_{i=0}^{t} \binom{n}{i},$$

and we shall treat the cases of odd and even d separately. For odd $d, t = \frac{d-1}{2}$ so

$$2^{d-1} > \sum_{i=0}^{\frac{1}{2}(d-1)} \binom{n}{i} \tag{4.1}$$

Now $2^d = (1+1)^d = \sum_{i=0}^{d} \binom{d}{i}$, and because of the symmetry property of the binomial coefficients, $\binom{d}{i} = \binom{d}{d-i}$, the first half of the terms in this sum of $d+1$ terms, those from $i = 0$ to $i = \frac{1}{2}(d-1)$, are identical to the remaining terms from $i = \frac{1}{2}(d+1)$ to $i = d$.

So

$$2^d = 2 \times \sum_{i=0}^{\frac{1}{2}(d-1)} \binom{d}{i},$$

so inequality (4.1) becomes

$$\sum_{i=0}^{\frac{1}{2}(d-1)} \binom{d}{i} > \sum_{i=0}^{\frac{1}{2}(d-1)} \binom{n}{i}$$

which can only be true if $d > n$, which is clearly impossible!

Moving on to the case of even $d, t = \frac{d-2}{2}$ so (4.1) is replaced by

$$2^{d-1} > \sum_{i=0}^{\frac{1}{2}(d-2)} \binom{n}{i} \tag{4.2}$$

You should check that when $d = 2$ (4.2) holds for all $n \geq 2$; when $d = 4$, only for $n = 4, 5$ and 6; for $d = 6$, only for $n = 6$ and 7; for $d = 8$, only for $n = 8$.

For general even d, using the same trick as for odd d, $2^d = \sum_{i=0}^{d} \binom{d}{i}$, which we now split up as

$$\sum_{i=0}^{\frac{1}{2}(d-2)} \binom{d}{i} + \binom{d}{\frac{d}{2}} + \sum_{i=\frac{1}{2}(d+2)}^{d} \binom{d}{i}$$

$$= 2 \sum_{i=0}^{\frac{1}{2}(d-2)} \binom{d}{i} + \binom{d}{\frac{d}{2}}$$

So

$$2^{d-1} = \sum_{i=0}^{\frac{1}{2}(d-2)} \binom{d}{i} + \frac{1}{2}\binom{d}{\frac{d}{2}}.$$

Equation (4.2) then becomes the requirement that

$$\sum_{i=0}^{\frac{1}{2}(d-2)} \binom{d}{i} + \frac{1}{2}\binom{d}{\frac{d}{2}} > \sum_{i=0}^{\frac{1}{2}(d-2)} \binom{n}{i} \tag{4.3}$$

Notice that the left hand side of this is independent of n whereas the right hand side is an increasing function of n. This gives us the useful result that:

> If, for a given value of d, (4.3) fails for some value of n, then it must also fail for all larger values of n. (4.4)

Also, $n \geq d$ for *any* code, and it is easy to see that (4.3) *is* satisfied for $n = d$. Before getting too excited by this result notice that binary codes with $n = d$ are somewhat trivial; they only contain two codewords. (Just apply the Singleton bound or use common sense!)

For $n = d + 1$ (4.3) becomes

$$\sum_{i=0}^{\frac{1}{2}(d-2)} \binom{d}{i} + \frac{1}{2}\binom{d}{\frac{d}{2}} > \sum_{i=0}^{\frac{1}{2}(d-2)} \binom{d+1}{i},$$

which it is convenient to rearrange as

$$\sum_{i=0}^{\frac{1}{2}(d-2)} \left[\binom{d+1}{i} - \binom{d}{i}\right] < \frac{1}{2}\binom{d}{\frac{d}{2}}. \tag{4.5}$$

If $d = 2$ this reduces to the obviously true $0 < \frac{1}{2}\binom{2}{1}$. For $d \geq 4$, if we note that the $i = 0$ term is zero, the left hand sum is

$$\sum_{i=1}^{\frac{1}{2}(d-2)} \left[\binom{d+1}{i} - \binom{d}{i}\right],$$

which, by the Pascal triangle recurrence relation is

$$\sum_{i=1}^{\frac{1}{2}(d-2)} \binom{d}{i-1},$$

then (4.5) reduces to

$$\sum_{i=1}^{\frac{1}{2}(d-2)} \binom{d}{i-1} < \frac{1}{2}\binom{d}{\frac{d}{2}}. \tag{4.6}$$

The arithmetic you did following inequality (4.2) has already confirmed that (4.6) holds for $d = 4$ and 6, but not 8. Before going on make one more preliminary calculation that (4.6) fails for $d = 10$ too. [Ex 4]

To make further progress we take just the last term, $\binom{d}{\frac{d}{2}-2}$, of the sum of positive terms on the left hand side of (4.6) and compare it with the right hand side. Specifically, we form the ratio of these two numbers,

$$\binom{d}{\frac{d}{2}-2} \div \frac{1}{2}\binom{d}{\frac{d}{2}}.$$

This simplifies to $\dfrac{2d}{d+4} \times \dfrac{d-2}{d+2}$ which is clearly as increasing function of d, and is greater than 1 when $d = 12$. So for all $d \geq 12$ just one term of the left hand side of (4.6) is already greater than the right hand side. Hence (4.6) fails for all $d \geq 12$.

Collecting all these results together, including (4.6), we obtain

Theorem 4.1 The only binary codes for which the Singleton bound is more powerful than the Hamming bound are:

1. $d = 2$, all n;
2. $d = 4, n = 4, 5, 6$;
3. $d = 6, n = 6, 7$;
4. d even $\geq 8, n = d$. □

Looking back at the proofs of the two bounds we see that to establish the Hamming bound we made direct use of the error correcting capability of the code whereas the Singleton derivation only used the minimum distance. This enables us to account for part 1 of Theorem 4.1, since these codes have a non-trivial minimum distance, but no error-correcting capability. Hence it is not surprising that the Singleton bound provides more information. As for the other cases, we shall see shortly that they are all rather uninteresting codes – none of them have more than four codewords. So for binary codes of any practical interest, forget the Singleton bound!

To find non-binary codes for which the Singleton bound wins we would have to solve

$$q^n \left[\sum_{i=0}^{t} \binom{n}{i} (q-1)^i \right]^{-1} > q^{n-d+1},$$

which amount to

$$q^{d-1} > \sum_{i=0}^{t} \binom{n}{i} (q-1)^i \tag{4.7}$$

and it is not difficult to find non-trivial codes which satisfy this. [Ex 5]

You may be interested in pursuing this investigation systematically for non-binary codes. The mathematics becomes rather messy and it is probably best to regard it as a computing project.

To end this section we present an argument which gives a *lower* bound for the best possible M: consider all $q - ary$ words of length n and choose one arbitrarily; then pick another, subject only to the restriction that its distance from the first is at least d; then another with distance $\geq d$ from the first two; ... and so on. At each stage of the process the next word can be freely chosen from all those words not in any of the spheres of radius $d-1$ centred on words already selected. Hence the process stops when there are no such words left. When this happens suppose M words have been selected and these constitute the set C. Then C is clearly a $q - ary$ (n, M, d) code. To estimate M note that the spheres $S(\mathbf{c}, d-1)$ centred on codewords of C must together contain all q^n words. [Ex 6]

Let U be the union of all these spheres, so we have

$$|U| = q^n \tag{4.8}$$

But of course, in general, a word will belong to more than one of the spheres, so if each sphere contains a words, then

$$Ma \geq |U| \tag{4.9}$$

By a similar calculation to that done in deriving the Hamming bound

$$a = \sum_{i=0}^{d-1} \binom{n}{i} (q-1)^i \tag{4.10}$$

Putting (4.8), (4.9) and (4.10) together we obtain:

Result 4.3 The Gilbert–Varshamov bound.

> For all n and q, and all $d \leq n$, there exists a code with
> $$M \geq q^n \left[\sum_{i=0}^{d-1} \binom{n}{i} (q-1)^i \right]^{-1}$$

[Ex 7]

4.3 Equivalent codes

This chapter has been concerned with finding out as much as we could concerning the number $A_q(n, d)$ which is defined as the largest M for which a $q - ary$ (n, M, d) code exists. In Exercise 7, for example, you found by applying the Hamming and Gilbert–Varshamov bounds that

$$5 \leq A_3(5, 3) \leq 22,$$

and then by explicitly constructing a ternary $(5, 6, 3)$ code, this was improved to

$$6 \leq A_3(5, 3) \leq 22.$$

If you wished to improve this further by the same method the next task would be to construct a ternary (5,7,3) code. Just to get a feel for the size of the problem notice that there are $3^5 = 243$ words of length 5. We need a 7 word subset of these, having a minimum distance of 3 or more. There are $\binom{243}{7}$ subsets, a number of the order of 9 billion, so even with a computer slave, doing a search of all the subsets is not a viable method.

An idea which gives more hope of progress in problems of this type is that there is no significant difference between many of the subsets. The sense in which this is true is illustrated by the three small codes listed below:

$$\begin{aligned} C &= \{abccba, bbcbbc, ccbaca\} \\ C' &= \{abccba, cbcbbb, acbacc\} \\ C'' &= \{bbcaba, abccbc, ccbbca\} \end{aligned}$$

C' has been obtained from C by re-ordering the symbols. In fact, C' is just the result of switching the first and last symbol in each codeword of C. C'' has been obtained from C by making some symbol changes in some of the positions. In fact a has been replaced by b and b by a in position 1, and in position 4 a has been replaced by b, b by c and c by a.

Now look at the first and last words of C. They differ in position 1 but not in position 6. The corresponding words in C' differ in position 6 but not in position 1, as should be obvious in view of the transformation actually used to get C' from C. This example can be generalized to give the following result: suppose C' is obtained from C by re-ordering the symbols of the C words in some way (the same re-ordering for all the words), so that the symbols in position i of the C words end up in position $\pi(i)$ in the C' words. Then the distance between any pair of C words must be the same as the distance between the corresponding C' words, because words of C differ in position i if and only if the corresponding words of C' differ in position $\pi(i)$. All that has happened is that the agreements and disagreements between pairs of words have moved to different positions; the number of each remains unchanged. An obvious consequence of this is

Theorem 4.2 Performing a positional permutation on the words of a code does not change its minimum distance. □

It is of course understood that the *same* permutation is applied to each codeword – otherwise the theorem is obviously false!

Next consider the transition, $C \to C''$. There we did not change any positions, but instead changed the symbols living in some of those positions. The vital feature of the changes made is that in any given position distinct symbols are replaced by distinct symbols. So, for example, in position 4 a, b, c were replaced by b, c, a respectively. We would disallow replacing a by b and replacing both b and c by a. The reason for making such a restriction is that we want the original code and the new code to have the following property: if in a given position two words of C differ, then the corresponding words of C'' also differ, and if in a given position two words of C agree, they will also agree in that position in C''. Clearly, without the restriction this property could fail. So the sort of symbol changes we do allow are *permutations* of the alphabet. It follows that the distance between a pair of words in C is not changed by doing symbol permutations in some or all of the positions. In particular we have

Theorem 4.3 If code C'' is produced from code C by performing symbol permutations at some or all of the positions of C, then $d(C'') = d(C)$. □

And of course it is *not* necessary to do the *same* permutation at each of the positions.

These two theorems motivate the following terminology:

Definition 4.1 Two codes are said to be equivalent if one can be obtained from the other by a sequence of positional and/or symbol permutations. [Ex 8, 9]

For the next part of the investigation it is convenient to refer to the *weight* of a word. We assume from now on that when the alphabet size is q the symbols are $0, 1, 2, \ldots, q-1$.

Definition 4.2 The weight of the word $\boldsymbol{x}$ is the number of positions not occupied by 0. We denote this by $w(\boldsymbol{x})$.

We illustrate the use of code equivalence by working out the exact value of $A_2(9, 6)$. Notice first that the bounds we have available tell us only that

$$1 \leq A_2(9, 6) \leq 11$$

and the lower bound is particularly uninformative! So let us try to find a best code C. For a binary code there are only two possible symbol permutations, $\binom{0\ 1}{0\ 1}$ and $\binom{0\ 1}{1\ 0}$. That is, we either make no change or we switch 0s and 1s. Suppose C is a best code and select any of its words, say $a_1a_2 \ldots a_9$. In each position in which $a_i = 1$ do the $0 \leftrightarrow 1$ switch. This produces an equivalent code C' in which one word is 000000000, and all the other words must have been transformed into words of weight 6 or more because their distances from 000000000 are at least 6. Now you can check that if $\boldsymbol{x}, \boldsymbol{y}$ are any two words with $w(\boldsymbol{x}) \geq 6$ and $w(\boldsymbol{y}) \geq 7$, they must agree in at least 4 of the 9 places, so $d(\boldsymbol{x}, \boldsymbol{y})$ would be ≤ 5. So if C' contains $\boldsymbol{y}$ with $w(\boldsymbol{y}) \geq 7$, then C' can only contain $\boldsymbol{o}$ and $\boldsymbol{y}$. So the only hope of doing better than a two word code is to have C' with all its non-zero words of weight 6.

So C' consists of $\boldsymbol{c}_1' = \boldsymbol{o}, \boldsymbol{c}_2'$ with $w(\boldsymbol{c}_2') = 6$ and possibly other words of weight 6. We can now do a positional permutation to bring all the 1s of $\boldsymbol{c}_2'$ to the left hand end. This does not of course affect $\boldsymbol{c}_1'$, because all its symbols are 0 anyway, nor does it affect the weight of any word. So the new code C'' looks like

$$\begin{aligned} \boldsymbol{c}_1'' &= 000000000 \\ \boldsymbol{c}_2'' &= 111111000 \\ \boldsymbol{c}_3'' &= \quad ? \\ &\;\;\vdots \end{aligned}$$

Since $d(\boldsymbol{c}_2'', \boldsymbol{c}_3'') \geq 6$ and $w(\boldsymbol{c}_3'') = 6, \boldsymbol{c}_3''$ must have three of its six 1s in the last three places, and of course this same argument holds for any other words which C'' may have. We can do a further positional permutation, not involving the last three places, which will bring the remaining three 1s of $\boldsymbol{c}_3''$ to the left hand end, without affecting $\boldsymbol{c}_1''$ or $\boldsymbol{c}_2''$. The new code C''' is

$$\begin{aligned} \boldsymbol{c}_1''' &= 000000000 \\ \boldsymbol{c}_2''' &= 111111000 \\ \boldsymbol{c}_3''' &= 111000111 \\ \boldsymbol{c}_4''' &= \quad ? \quad 111 \\ &\;\;\vdots \end{aligned}$$

You should now find it easy to convince yourself that to maintain $d(\boldsymbol{c}_2, \boldsymbol{c}_i) \geq 6$ and $d(\boldsymbol{c}_3, \boldsymbol{c}_i) \geq 6$ for $i \geq 4$, the remaining three 1s of $\boldsymbol{c}_i$ have to go in positions 4, 5 and 6. This forces the conclusion that $\boldsymbol{c}_4'''$ has to be 000111111, and that there can be no other words.

Hence $A_2(9,6) = 4$, and if you look back over this argument you will see that it proves more: *all* binary (9,4,6) codes are equivalent to the one just constructed. [Ex 10, 11]

Putting the results of the last two theorems together it is clear that equivalent codes have the same minimum distance. The converse is false, however: it is not difficult to find pairs of (n, M, d) codes which are demonstrably not equivalent. [Ex 12]

So we now know that if two codes are equivalent one is t-error correcting if and only if the other is. But this does not tell the whole story. Suppose C, C' are both 1-error correcting. This only tells us that both correct all instances of words received with a single error. It says nothing about how they perform with two or more errors. In order to define an overall error processing performance we consider a complete decoding scheme. That is, every received word is actually interpreted as some code word, so the option of declaring a received word to be uncorrectable, because it is equally (and minimally) distant from several codewords, is not available.

A sensible nearest neighbour decoding scheme under these circumstances is as follows:

Let $\boldsymbol{c}$ be sent and $\boldsymbol{r}$ received.

1. Find the codeword (or codewords) whose distance from $\boldsymbol{r}$ is minimal.

2.(a) If there is just one such codeword decode $\boldsymbol{r}$ as this codeword.

(b) If there are k codewords at this minimal distance from $\boldsymbol{r}$, chose one 'at random' and decode $\boldsymbol{r}$ to this.

So in case (a), if the unique codeword is $\boldsymbol{c}$ the probability of correct decoding is 1, and if not, the probability is 0. In case (b), if $\boldsymbol{c}$ is one of the k codewords then decoding will be correct with probability $\frac{1}{k}$, and if not, 0.

Now imagine performing the transformation τ (symbol permutations + a positional permutation) which takes C to the equivalent code C', but do τ not just on the codewords but on all words. This sets up a one-to-one correspondence on the set of all words, and it follows from our discussion prior to Theorems 4.2 and 4.3 that τ preserves distances. That is, for all words, $\boldsymbol{x}$, $\boldsymbol{y}$, $d(\boldsymbol{x}, \boldsymbol{y}) = d(\tau(\boldsymbol{x}), \tau(\boldsymbol{y}))$. From this and the complete decoding algorithm we have:

Theorem 4.4 If C, C' are equivalent codes then a received word $\boldsymbol{r}$ is correctly decoded in C with probability p if and only if $\tau(\boldsymbol{r})$ is correctly decoded in C' with probability p. □

It is in this sense that equivalent codes are identical as far as error correcting is concerned.

4.4 Distance isomorphic codes

Consider the codes C, C' with the following codewords.

$$\begin{aligned} C &= \{c_1, c_2, c_3, c_4, c_5\} \\ &= \{0111010101, 1011101110, 1011011101, \\ &\quad 0000111100, 1101101101\} \end{aligned}$$

$$\begin{aligned} C' &= \{c'_1, c'_2, c'_3, c'_4, c'_5\} \\ &= \{0110000000, 1111100110, 1000000000, \\ &\quad 1111010001, 0111111000\} \end{aligned}$$

The tables below show the distances between pairs of words in each code (omitting the entries for $d(\boldsymbol{c}_i, \boldsymbol{c}_i) = 0$).

C	$\boldsymbol{c}_1$	$\boldsymbol{c}_2$	$\boldsymbol{c}_3$	$\boldsymbol{c}_4$	$\boldsymbol{c}_5$
$\boldsymbol{c}_1$		7	3	6	5
$\boldsymbol{c}_2$			4	5	4
$\boldsymbol{c}_3$				5	4
$\boldsymbol{c}_4$					5

C'	$\boldsymbol{c}'_1$	$\boldsymbol{c}'_2$	$\boldsymbol{c}'_3$	$\boldsymbol{c}'_4$	$\boldsymbol{c}'_5$
$\boldsymbol{c}'_1$		5	3	4	4
$\boldsymbol{c}'_2$			6	5	5
$\boldsymbol{c}'_3$				5	7
$\boldsymbol{c}'_4$					4

The numbers in the two tables are identical apart from the order in which they appear. Indeed, in this case we can change the order in which the words of C' are displayed so that the tables are absolutely identical, as shown below.

C'	$\boldsymbol{c}'_3$	$\boldsymbol{c}'_5$	$\boldsymbol{c}'_1$	$\boldsymbol{c}'_2$	$\boldsymbol{c}'_4$
$\boldsymbol{c}'_3$		7	3	6	5
$\boldsymbol{c}'_5$			4	5	4
$\boldsymbol{c}'_1$				5	4
$\boldsymbol{c}'_2$					5

[Ex 13]

This leads naturally to:

Definition 4.3 Two $q-ary$ codes C, C' of the same length and size are called *distance isomorphic* if their words can be ordered so that for all $\boldsymbol{c}_i, \boldsymbol{c}_j \in C$ and all $\boldsymbol{c}'_i, \boldsymbol{c}'_j \in C'$, $d(\boldsymbol{c}_i, \boldsymbol{c}_j) = d(\boldsymbol{c}'_i, \boldsymbol{c}'_j)$.

You have seen that equivalent codes are necessarily distance isomorphic, and Exercise 13 has shown the converse to be false. A natural question is whether distance isomorphic codes are necessarily equally good error correctors. To answer this take the simpler pair of codes,

$$\begin{aligned} C &= \{0000, 0011, 0101, 0110\} \\ C' &= \{0111, 1011, 1101, 1110\} \end{aligned}$$

It is easy to check that they are distance isomorphic but not equivalent. To see that they are also not identical error correctors consider the word $\boldsymbol{r} = 1000$ which has distance 1 from the first codeword of C and 3 from the rest. Then check that of all the words at distance 1 from any codeword of C', none of them have distance 3 from all the rest. Furthermore, if complete probabilistic decoding is used, then by considering the outcome for each of the sixteen (transmitted codeword, received word) pairs where the received words contain a single error, you can check that the probability of such a word being correctly decoded in C is $\frac{1}{2}$, but in C' only $\frac{7}{16}$. [Ex 14]

4.5 Geometry and Hamming space

Up until now we have stressed the similarity of Hamming distance between words and ordinary Euclidean distance between points. This similarity stems from the fact that both satisfy the three defining properties of a metric (see Chapter 2). It enabled us to use geometric language and analogy to assist in understanding various consequences of the triangle inequality, sphere-packing, etc. What is *not* often emphasized is that there are also important *differences* between Euclidean space and Hamming space, and the discussion in the previous section can highlight one of these.

Think first of ordinary 2-dimensional (plane) geometry. An *isometry* of the plane is a mapping f of the set of points of the plane to itself, with the property of being distance preserving. That is, if P, Q are any two points and d is ordinary Euclidean distance, then

$$d(P,Q) = d(f(P), f(Q)).$$

Now if $S = \{P_1, P_2, \ldots, P_k\}$ is any set of points in the plane and $S' = \{P'_1, P'_2, \ldots, P'_k\}$ is another with the property that, for all $i, j, d(P_i, P_j) = d(P'_i, P'_j)$, then there is an isometry *of the whole plane* which maps P_1 to P'_1, P_2 to $P'_2, \ldots$ etc. In the language of classical geometry we can say that if we know the distances between every pair of points in some geometric configuration S then we know everything about its shape and size – only its position is unknown. So any two configurations S, S' with the same set of distances are *congruent*. Furthermore, if P is any point and its distances from the points of S are $d_1, d_2, \ldots, d_k$, then there will be some point P' with the same sequence of distances from the corresponding points of S'. This is true of Euclidean geometry of all dimensions, but it is a property which fails for Hamming space. You have already seen a counter-example in the codes of Exercise 14.

We define an isometry of Hamming space Z_q^n in the same way as for Euclidean space – just replace geometric distance by Hamming distance.

So when we compare the notion of equivalence and the weaker notion of distance isomorphism we see that the latter is just not strong enough to guarantee identical error correcting performances of the codes. Is this why equivalence is defined the way it is? Could there be some relationship between codes, stronger than distance isomorphism but weaker than equivalence, which still ensures that the codes are identical error correctors? In order to pursue this further it is convenient to make another definition. We have been thinking of codes as *sets* of words, but in our discussion of distance isomorphism the codes were presented in a specific *order*, $\mathbf{c}_1, \mathbf{c}_2, \ldots, \mathbf{c}_M$. Now suppose C and C' are codes such that positional and symbol permutations on C' will transform it into C'' which is identical (as a set) to C. However, if C is large it may not be easy to recognise that C and C'' are the same set because the order of the words of C'' may be jumbled in a fairly complicated way relative to C. Partly for this reason we define the notion of *strict* equivalence between *ordered* codes.

Definition 4.4 Two ordered codes $\langle \boldsymbol{c}_1, \boldsymbol{c}_2, \ldots, \boldsymbol{c}_M \rangle$ and $\langle \boldsymbol{c}'_1, \boldsymbol{c}'_2, \ldots, \boldsymbol{c}'_M \rangle$ are strictly equivalent if one can be obtained from the other using symbol and positional permutations and no re-ordering of the words.

The following two theorems settle our questions by showing that codes with identical error correcting performance must be equivalent.

Theorem 4.5 Let $f : Z_2^n \to Z_2^n$ be an isometry. Choose an ordering $\langle \boldsymbol{x}_1, \ldots, \boldsymbol{x}_{2^n} \rangle$ of Z_2^n. Then the list $L = \langle f(\boldsymbol{x}_1), \ldots, f(\boldsymbol{x}_{2^n}) \rangle$ is strictly equivalent to a list $L' = \langle f(\boldsymbol{x}_1)', \ldots, f(\boldsymbol{x}_{2^n})' \rangle$ in which, for all $i, \boldsymbol{x}_i$ and $f(\boldsymbol{x}_i)'$ agree in their first symbols.

Preliminary note: it is convenient to take $\boldsymbol{x}_1, \ldots, \boldsymbol{x}_{2^{n-1}}$ to be the words starting with 0, and let these words be listed in order of increasing weight, and list the words beginning with 1 in order of increasing weight too.

Proof. Let Z_2^n be listed as above. Then do the following equivalence transformations on L.

1. Do a symbol permutation on each of the positions in which $f(\boldsymbol{x}_1)$ has a 1. Call this transformation π. Then $\pi(f(\boldsymbol{x}_1)) = \boldsymbol{0}$, and by considering distances of the other words from $\boldsymbol{0}$ we see that $w(\boldsymbol{x}_i) = w(\pi(f(\boldsymbol{x}_1)))$ for all i. (1)
2. Then do a positional permutation σ so that the words of weight 0 and 1 in L and in the new list coincide. That is, for all $\boldsymbol{x}_i$ of weight 0 or 1, $\boldsymbol{x}_i = \sigma\pi f(\boldsymbol{x}_i)$. (2)

Claim. $\langle \sigma\pi f(\boldsymbol{x}_i) : i = 1, 2, \ldots, 2^n \rangle$ is the required list, L'. To show this, all we have to do is demonstrate that no word in the first 2^{n-1} words of $\langle \sigma\pi f(\boldsymbol{x}_i) \rangle$ can start with a 1. Suppose this is false. That is, there is some word $\boldsymbol{x}_j, j \leq 2^{n-1}$, for which $\sigma\pi f(\boldsymbol{x}_j)$ starts with 1. Notice that since positional permutations do not change weights $w(\pi f(\boldsymbol{x}_i)) = w(\sigma\pi f(\boldsymbol{x}_i))$ for all i, and hence by (1) above, $w(\boldsymbol{x}_i) = w(\sigma\pi f(\boldsymbol{x}_i))$ for all i.

So by (2), $\boldsymbol{x}_j$ must have weight ≥ 2. Let $w = w(\sigma\pi f(\boldsymbol{x}_j))$ and consider the distances of $\boldsymbol{x}_j$ and of $\sigma\pi f(\boldsymbol{x}_j)$ from each of the words $\boldsymbol{u}$ of weight 1 (the 'unit' words). $d(\boldsymbol{x}_j, \boldsymbol{u}) = w - 1$ for exactly w of the unit words, but $d(\sigma\pi f(\boldsymbol{x}_j), \boldsymbol{u}) = w - 1$ for exactly $w - 1$ of the unit words so the map $\boldsymbol{x} \mapsto \sigma\pi f(\boldsymbol{x})$ is *not* an isometry. This contradiction establishes the claim.

Theorem 4.6 Let $f : Z_2^n \mapsto Z_2^n$ be an isometry, with L being a listing of Z_2^n as in the previous theorem. Then $\langle f(\boldsymbol{x}_i) : i = 1, 2, \ldots, 2^n \rangle$ is strictly equivalent to L.

Proof. We use induction on n. The result is trivially true for $n = 1$, so suppose it is true for isometries of Z_2^n for all $n < k$.

Then let f be an isometry of Z_2^k.

First do the required transformations to convert $\langle f(\boldsymbol{x}_i)\rangle$ to L' as in the previous theorem. Then the function g which takes each member of L to the corresponding member of L' is also an isometry of Z_2^k, in which all words in the top halves of L and L' start with 0 and the words in the bottom halves of both lists start with 1.

Now delete the first bit from all words so that L and L' become new lists $\hat{L}$ and $\hat{L}'$ of words of length $k-1$.

Let $\hat{g}$ match up members of $\hat{L}$ to members of $\hat{L}'$ in the same way as g matched L to L', and let $\hat{g}_t$ and $\hat{g}_b$ be the restrictions of $\hat{g}$ to $\hat{L}_t$ and $\hat{L}_b$, the top and bottom halves of $\hat{L}$. Then clearly $\hat{g}_t$ and $\hat{g}_b$ are both isometries of Z_2^{k-1}.

By our inductive hypothesis $\hat{L}_t$ and $\hat{L}'_t$ are strictly equivalent, and because both lists start with **0** no symbol permutations are necessary to convert $\hat{L}'_t$ to $\hat{L}_t$. So let the positional permutation required to convert $\hat{L}'_t$ to $\hat{L}_t$ be done, but do it to the whole of $\hat{L}'$ and let the result be $\hat{L}''$.

Then $\hat{L}_t = \hat{L}''_t$ and $\hat{L}_b \to \hat{L}''_b$ is an isometry. We can think of this positional permutation as being done to the unshortened (length k) words – the permutation just happens not to move the first bits – so we also have $L_t = L''_t$ and $L_b \to L''_b$ is an isometry.

To complete the proof we show that in fact we have forced $L_b = L''_b$.

To this end, let $\boldsymbol{x}_i = 1a_2a_3\ldots a_k$ and $\boldsymbol{y}_i = 1b_2b_3\ldots b_k$ be corresponding words in L_b, L''_b respectively. Consider the word $\boldsymbol{x}_j = 0a_2a_3\ldots a_k$ in L_t (which is matched with the same word in L''_t.

So we have the following situation:

$$\begin{array}{ccc} L & & L'' \\ \vdots & & \vdots \\ \vdots & & \vdots \\ \boldsymbol{x}_j & \longrightarrow & \boldsymbol{x}_j \\ \vdots & & \vdots \\ \boldsymbol{x}_i & \longrightarrow & \boldsymbol{y}_i \\ \vdots & & \vdots \end{array}$$

Because this matching is an isometry on Z_2^k we have

$$\begin{aligned} d(\boldsymbol{x}_j, \boldsymbol{x}_i) &= d(\boldsymbol{x}_j, \boldsymbol{y}_i) \\ \text{i.e.} \qquad 1 &= 1 + d(a_2a_3\ldots a_k, b_2b_3\ldots b_k) \end{aligned}$$

which can only be true if $\boldsymbol{x}_i = \boldsymbol{y}_i$. □

4.6 Perfect codes

Earlier in this Chapter we found that the binary Hamming (7, 16, 3) code was a best possible code in the sense that 16 is the upper limit on M imposed by the sphere packing bound for all binary codes with $n = 7, d = 3$, and that the Hamming code actually achieves this limit. Such codes are called *perfect* codes and much research effort has gone into looking for all of these (rather rare) codes. Here is the formal definition:

Definition 4.5 A q-ary $(n, M, d)t$-error correcting code is called perfect if

$$M = q^n \left[\sum_{i=0}^{t} \binom{n}{i} (q-1)^i \right]^{-1}$$

If you look back at the proof of the sphere-packing bound in section 4.2 you will see that perfect codes can be neatly characterized geometrically as follows:

Theorem 4.7 The t-error correcting code C is perfect if and only if the set of spheres, $\{S(\mathbf{c}, t) : \mathbf{c} \in C\}$ is pairwise disjoint and their union is the set of all words. □

[Ex 15]

Perfect codes also have a neat, clear-cut error-correcting property which you are asked to prove as the next exercise.

Theorem 4.8 A t-error correcting perfect code will not correctly decode any received word with more than t errors. □

[Ex 16]

For a general (non-perfect) t-error correcting code the spheres of radius t centred on codewords do not cover all words. Just how much would these spheres have to be enlarged so that they do? Specifically, if C is a code, what is the smallest value of r, say ρ such that the family of spheres $\{s(\mathbf{c}, r) : \mathbf{c} \in C\}$ covers all words? ρ is called the *covering radius* of C. It is the smallest r such that every word is within distance r of some codeword. Clearly $\rho \geq t$ and a code is perfect if and only if $\rho = t$.

Determining ρ for specific codes and finding bounds on ρ are important (hard) combinatorial problems, and we shall content ourselves with scratching the surface.

Suppose we have an (n, M, d) code which is optimal in the sense that no further codewords may be added to it without decreasing d. Then by the argument which led to the Gilbert–Varshamov bound we know that the union of codeword centred spheres of radius $d - 1$ covers all words.

But ρ was *defined* as the smallest radius for which this is true, so we have proved

Theorem 4.9 For t-error correcting codes with covering radius ρ and minimum distance $d, t \leq \rho$, and if the code is optimal $\rho \leq d-1$. □

We now determine the covering radius of the Hamming 8-bit code, and this will also introduce some important ideas for later use. Previous exercises have suggested that this code has a minimum distance of 4, but no proof of this has been given. To rectify this omission we first establish a useful property of Hamming distance and weight for binary codes:

Theorem 4.10 For any two binary words $\boldsymbol{x}, \boldsymbol{y}, d(\boldsymbol{x}, \boldsymbol{y}) = w(\boldsymbol{x}+\boldsymbol{y}) = w(\boldsymbol{x}) + w(\boldsymbol{y}) - 2w(\boldsymbol{x} \odot \boldsymbol{y})$, where the 'sum' and 'product' words $\boldsymbol{x}+\boldsymbol{y}$ and $\boldsymbol{x} \odot \boldsymbol{y}$ are formed bit-wise from $\boldsymbol{x}$ and $\boldsymbol{y}$, with the arithmetic done modulo 2.

[For example, if $\boldsymbol{x} = 01110101$ and $\boldsymbol{y} = 01011110$, then $\boldsymbol{x}+\boldsymbol{y} = 00101011$ and $\boldsymbol{x} \odot \boldsymbol{y} = 01010100$].

Proof. The first equality is clear because the bits which contribute 1 to $d(\boldsymbol{x}, \boldsymbol{y})$ are just those where $\boldsymbol{x}$ and $\boldsymbol{y}$ differ, and these are precisely the bits of $\boldsymbol{x}+\boldsymbol{y}$ which are 1.

The second equality can also be established by considering the contribution from individual bits. For example, if the ith bits of $\boldsymbol{x}$ and $\boldsymbol{y}$ are both 1 the contribution to $w(\boldsymbol{x}+\boldsymbol{y})$ is 0, and the contribution to $w(\boldsymbol{x}) + w(\boldsymbol{y}) - 2w(\boldsymbol{x} \odot \boldsymbol{y})$ is $1+1-2(1.1) = 0$. You can check the other cases easily, but remember to interpret the arithmetic correctly, as shown below.

$$
\begin{array}{lcccl}
\text{bitwise mod2 addition} & & & & \text{bitwise mod 2 multiplication} \\
\quad\uparrow & & & & \quad\uparrow \\
w(\boldsymbol{x}+\boldsymbol{y}) = w(\boldsymbol{x}) & + & w(\boldsymbol{y}) & - & 2w(\boldsymbol{x} \odot \boldsymbol{y}) \\
 & \uparrow & & \uparrow & \nwarrow \\
 & \text{ordinary addition,} & & \text{subtraction,} & \text{multiplication}
\end{array}
$$

□

[Ex 17]

Theorem 4.11 The Hamming 8-bit code has covering radius 2.

Proof. Let C, C' be the Hamming 8- and 7-bit codes respectively. We know $d(C') = 3$, and since the codewords of C are just those of C' with one extra bit added, $d(C)$ can only be 3 or 4.

If $\boldsymbol{x}, \boldsymbol{y} \in C$, then $w(\boldsymbol{x}), w(\boldsymbol{y})$ are both even, and $2w(\boldsymbol{x} \odot \boldsymbol{y})$ is necessarily even, so by Theorem 4.10 $d(\boldsymbol{x}, \boldsymbol{y})$ is even. So $d(C)$ must be 4.

Furthermore, C is optimal since if not there would be a word $\boldsymbol{x}$ with $d(\boldsymbol{x}, \boldsymbol{c}) \geq 4$ for all $\boldsymbol{c}$ in C, and this would mean that $\boldsymbol{x}'$, the word consisting of the first seven bits of $\boldsymbol{x}$, would be at distance at least 3 from all codewords of C'. This is impossible since we know C' is optimal. Also C is not perfect, so applying Theorem 4.9 we have $1 < \rho \leq 3$. To settle it for 2, let $x_1x_2 \dots x_8$ be any 8-bit word. We know that $x_1x_2 \dots x_7$ is at distance 0 or 1 from some (unique) codeword $\mathbf{c}'(= c_1c_2\ldots c_7)$ of C' since C' is one-error correcting and perfect. Let $c_1c_2\ldots c_8$ be the corresponding codeword of C. Then

$$d(x_1x_2\ldots x_8, c_1c_2\ldots c_8) = \begin{cases} d(x_1x_2\ldots x_7, c_1c_2\ldots c_7) & \text{if } x_8 = c_8 \\ d(x_1x_2\ldots x_7, c_1c_2\ldots c_7) + 1 & \text{if } x_8 \neq c_8 \end{cases}$$
$$= 0 \text{ or } 1 \text{ or } 2$$

That is, every word in Z_2^8 is within distance 2 of a codeword of C, so $\rho(c) = 2$. □

[Ex 18]

Another use of Theorem 4.10 is to save some work in finding the numbers $A_2(n,d)$. Essentially, the result we now prove shows that for each $A_2(n,d)$ you calculate with an odd d, you can immediately write down another value of $A_2(n,d)$ with even d.

Theorem 4.12 For each odd d, $A_2(n,d) = A_2(n+1,d+1)$.

Proof. Let C be a binary (n,M,d) code with d odd. Construct the code D of length $n+1$ simply by adding an overall parity check bit to each codeword of C. This can only leave d unchanged or increase it by 1. To see that the latter always occurs consider all pairs of words of C for which $d(\boldsymbol{x},\boldsymbol{y}) = d$. By Theorem 4.10 $w(\boldsymbol{x})$ and $w(\boldsymbol{y})$ must have opposite parity, so the words of D constructed from $\boldsymbol{x}$ and $\boldsymbol{y}$ also differ in their last place, so D has minimum distance $d+1$.

Now let C' be a code with arbitrary minimum distance d'. Select any two of its codewords which differ in d' places and select one of these positions. Construct a new code D' by deleting the bits in this position from all codewords of C'. The result is clearly a code with $d(D') = d(C') - 1$. We have now proved that for each binary (n,M,d) code with d odd there is a binary $(n+1,M,d+1)$ code, and for each binary (n,M,d') code there is a binary $(n-1,M,d'-1)$ code.

Now let d be odd and suppose $A_2(n,d) = A$. Using what has just been proved we have

$$A_2(n,d) = A \Rightarrow \exists(n,A,d) \text{ code } \Rightarrow \exists(n+1,A,d+1) \text{ code } \Rightarrow$$
$$A_2(n+1,d+1) = A+k, k \geq 0,$$

and

$$A_2(n+1,d+1) = A+k \Rightarrow \exists(n+1,A+k,d+1) \text{ code } \Rightarrow$$
$$\exists(n,A+k,d) \text{ code } \Rightarrow A_2(n,d) \geq A+k$$

Hence $A_2(n,d) = A \Rightarrow A_2(n,d) = A+k$ with $k \geq 0$, so k can only be 0.

□

The Hamming 7-bit code is an example of a binary $(n,M,3)$ code with parameters of the form $n = 2^m - 1, M = 2^{n-m}$. In 1962 J L Vasil'ev found a method of generating a family of binary perfect 1-error correcting codes from such an $(n, M, 3)$ 'seed' code, and what follows is an adaptation of his method.

From the original $(n, M, 3)$ code C construct another code D of length $2n+1$ whose first n bits are any word $\boldsymbol{u}$ of Z_2^n, the next n bits are $\boldsymbol{u}+\boldsymbol{v}$ where $\boldsymbol{v}$ is any word of C, and the final bit was chosen in a special way by Vasil'ev for his purposes, but for us it suffices to let it be an overall parity check bit for $\boldsymbol{u}$.

A useful notation for D is

$$D = \{\boldsymbol{u}|\boldsymbol{u}+\boldsymbol{v}|f(\boldsymbol{u}) : \boldsymbol{u} \in Z_2^n, \boldsymbol{v} \in C\}$$

where $f(\boldsymbol{u}) = \begin{cases} 0 \text{ if } w(\boldsymbol{u}) \text{ is even} \\ 1 \text{ if } w(\boldsymbol{u}) \text{ is odd} \end{cases}$,

and $\boldsymbol{a}|\boldsymbol{b}$ just means the word formed by writing the bits of $\boldsymbol{b}$ immediately after the bits of $\boldsymbol{a}$, so you can read '|' as 'followed by'.

Theorem 4.13 If the seed code has parameters of the form $n = 2^m - 1, M = 2^{n-m}, d = 3$, then the Vasil'ev construction gives a perfect code with parameters of the same form, m being replaced by $m+1$.

Proof. This is carried out in the following exercises. □

[Ex 19, 20, 21, 22]

You will meet more perfect codes in later chapters.

4.7 The Plotkin bound

To end this chapter we return to our theme of upper bounds for $A_2(n, d)$. A rather loose connection exists between this and the previous section in that the $\boldsymbol{u}|\boldsymbol{u}+\boldsymbol{v}$ construction and the bound we are about to discuss are both attributed to M. Plotkin. The Plotkin construction will be used significantly again in Chapter 9.

One way of improving the bounds obtained so far is to be slightly less ambitious; instead of aiming to prove that *all* (n, M, d) codes have $M \leq f(n, d)$ for some function f, show that such a result holds for a restricted range of n and d values. The Plotkin bound is one such result, which holds for codes with $d > \frac{n}{2}$, so could be useful in investigating codes for very noisy channels where we require a large minimum distance compared with the word length.

The derivation of the bound is another combinatorial argument which uses the powerful technique of estimating the same quantity in two different ways and comparing the results. We also need a fact concerning the 'greatest integer function' which is the subject of the next exercise. [Ex 23]

Theorem 4.14 If C is a binary (n, M, d) code with $d > \frac{n}{2}$, then $M \leq \frac{2d}{2d-n}$.

Proof. Let $S = \sum d(\boldsymbol{u}, \boldsymbol{v})$ where the sum is taken over all M^2 ordered pairs $(\boldsymbol{u}, \boldsymbol{v})$ in $C \times C$. The M zero terms of this sum are just those in which $\boldsymbol{u} = \boldsymbol{v}$. The remaining $M^2 - M$ terms each contribute at least d to S.

Hence $S \geq (M^2 - M)d$, which gives a *lower* bound for S. But we can also get an *upper* bound for S as follows. Write out the codewords of C as a binary array:

$$\begin{aligned}
\mathbf{c}_1 &= c_{11}c_{12}\ldots c_{1n} \\
\mathbf{c}_2 &= c_{21}c_{22}\ldots c_{2n} \\
&\vdots \\
&\vdots \\
\mathbf{c}_M &= c_{M1}c_{M2}\ldots c_{Mn}
\end{aligned}$$

and consider the contribution to S from the kth column, which has z_k zeros and $M - z_k$ ones. (c_{ik}, c_{jk}) contributes 1 to S if $c_{ik} \neq c_{jk}$ and contributes nothing otherwise. There are $z_k(M - z_k)$ $(0,1)$ pairs and $(M - z_k)z_k$ $(1,0)$ pairs, so the total contribution of the column to S is $2z_k(M - z_k)$. Summing over all the columns,

$$S = \sum_{k=1}^{n} 2z_k(M - z_k). \qquad (1)$$

Now treating M as a constant and z_k as a real variable, $z_k(M - z_k)$ is a quadratic function of z_k whose graph is a parabola symmetric about the axis $z_k = \frac{M}{2}$ and peaking at this value of z_k.

$$\text{So} \quad S \leq 2n\frac{M}{2}\left(M - \frac{M}{2}\right) = \frac{nM^2}{2} \qquad (2)$$

Combining these two inequalities for S gives

$$(M^2 - M)d \leq \frac{nM^2}{2}$$

from which $2(M - 1)d \leq nM$. Rearranging, $M(2d - n) \leq 2d$, so, provided $2d - n > 0$,

$$M \leq \frac{2d}{2d - n}$$ □

[Ex 24, 25]

By using the fact that z_k and M are actually *integers* we can achieve a slight improvement : the Plotkin bound above becomes

$$\begin{aligned}
M &\leq \left\lfloor \frac{2d}{2d - n} \right\rfloor \\
&\leq 2\left\lfloor \frac{d}{2d - n} \right\rfloor + 1 \text{ by Exercise 23.}
\end{aligned}$$

But this last expression is odd, so *when M is even* we must have

$$M \leq 2\left[\frac{d}{2d-1}\right]. \qquad (3)$$

When M is odd $z_k(M - z_k)$ is maximized by $z_k = \dfrac{M \pm 1}{2}$, in which case (3) is replaced by

$$S \leq 2n\left(\frac{M-1}{2}\right)\left(\frac{M+1}{2}\right) = \frac{N(M-1)(M+1)}{2} \qquad (4)$$

Combining this with $S \geq (M^2 - M)d$ we have

$$(M^2 - M)d \leq \frac{n(M-1)(M+1)}{2}$$

which rearranges to

$$M \leq \frac{n}{2d-n} = \frac{2d}{2d-n} - 1, \text{and since } M \text{ is an integer}$$

$$M \leq \left[\frac{2d}{2d-n}\right] - 1$$

$$\leq 2\left[\frac{d}{2d-n}\right] \text{ by Exercise 23 again.}$$

Putting the two cases together we see that $M \leq 2\left[\dfrac{d}{2d-n}\right]$ holds irrespective of whether M is odd or even.

The upper and lower bounds for S, from which the Plotkin bound was derived, are both relatively crude, so it is surprising how good the Plotkin bound actually is. For $n = 9, d = 6$ for example, it estimates:

$A_2(9,6) \leq \dfrac{12}{12-9} = 4$ and we found earlier in this chapter that the exact value of $A_2(9,6)$ is 4.

To see that our slight refinement can sometimes give a real improvement, consider the case $d = 10, n = 16$. The Plotkin bound gives $M \leq 5$, but the refinement improves this to $M \leq 4$.

4.8 Exercises for Chapter 4

1. Determine whether either of the two Hamming codes introduced in Chapter 2 are MDS codes.
2. Prove the geometric characterization of a t-error-correcting $q-ary$ code.
3. Can there be a ternary double-error-correcting code of length 10 containing at least 300 words?

4. Carry out the necessary checks of the various claims made by the following results: (4.2), (4.4) and (4.6).

5. Show that for any single-error-correcting code with a length not exceeding the alphabet size, the Singleton bound provides a tighter constraint than the Hamming bound for the size of the code.

6. Why do these spheres cover all words?

7. Using only the Hamming, Singleton and Gilbert–Varshamov bounds, what is the most which can be said about the best possible value of M for codes:

   (a) defined over Z_{10} with $n = 10, d = 5$;

   (b) defined over Z_3 with $n = 5, d = 3$;

   (c) defined over Z_3 with $n = 5, d = 4$?

   Explain why the Hamming bound gives the same result for (b) and (c).

   In (b) improve the lower bound for the best M by actually constructing a suitable code.

8. C is the code $\{aadcca, adcacd, cdabaa, dcbdbc\}$. By using symbol and/or positional permutations find an equivalent code C' with the following features:

   (a) Each word of C' starts with a different letter;

   (b) The first and last letters of each word of C' are the same;

   (c) b occurs twice in one position of the code.

   [The code C' is not unique!]

9. Let $\begin{pmatrix} s_1 & s_2 & s_3 & \dots & s_q \\ s'_1 & s'_2 & s'_3 & \dots & s'_q \end{pmatrix}$ denote the symbol permutation which, for each i, replaces symbol s_i by symbol s'_i, and let $\begin{pmatrix} 1 & 2 & 3 & \dots & n \\ 1' & 2' & 3' & \dots & n' \end{pmatrix}$ be the positional permutation which moves the symbol in position j to the new position j'.

   C is the code $\{bddac, abcda, abbbc, cdcdc, caddb, bccca\}$.

   Construct the equivalent code C_1 by applying the position permutation $\left(\begin{smallmatrix} 1 & 2 & 3 & 4 & 5 \\ 3 & 1 & 2 & 5 & 4 \end{smallmatrix}\right)$, then to the result of this apply the symbol permutation $\left(\begin{smallmatrix} a & b & c & d \\ a & d & b & c \end{smallmatrix}\right)$ in position 4.

   Then construct the equivalent code C_2 by applying the same two transformations to C, but in the reverse order.

10. Try to find $A_2(5,3)$ and (rather harder) $A_2(9,5)$ by using code equivalence.

11. If there is a binary (n, M, d) code, show that there is a binary $(n - 1, M', d')$ code with $M' \geq \frac{M}{2}$ and $d' \geq d$.

    [Hint: At least half the words of the (n, M, d) code must start with the same symbol. What is the result of deleting this symbol from just these words?]

    Deduce that $A_2(n, d) \leq 2A_2(n - 1, d)$.

12. Find such an example.

13. Why are the codes C, C' not equivalent?

14. Do the suggested checks.

15. Show that no perfect code can have an even minimum distance.

16. Prove Theorem 4.8.

17. Find a similar formula for that in Theorem 4.10 for ternary codes, of the form

    $$w(\boldsymbol{x} + \boldsymbol{y}) = w(\boldsymbol{x}) + w(\boldsymbol{y}) - f(\boldsymbol{x} \odot \boldsymbol{y})$$

    where $\odot$ and the first $+$ are modulo 3 multiplication and addition respectively, and f is a function to be found. [Hint: consider the number of 0s, 1s and 2s in $\boldsymbol{x} \odot \boldsymbol{y}$.]

18. Find an alternative proof that the 8-bit Hamming code has covering radius 2, based on the decoding algorithm given in Chapter 2.

19. Show that for any binary words $\boldsymbol{a}, \boldsymbol{b}, \boldsymbol{x}, \boldsymbol{y}$, $d(\boldsymbol{a}+\boldsymbol{b}, \boldsymbol{x}+\boldsymbol{y}) = d(\boldsymbol{a}+\boldsymbol{x}, \boldsymbol{b}+\boldsymbol{y})$.

20. Show that D has 2^{2n-m} codewords.

21. Show that $d(D) = 3$.

22. Show that D is perfect.

23. For any real number x, $[x]$ is called the greatest integer function of x, defined as the largest integer not larger than x. Prove that $2[x] \leq [2x] \leq 2[x] + 1$.

24. What information is obtained by applying the Plotkin bound argument when $d < \frac{n}{2}$?

25. The exact values of $A_2(n, 7)$ for $n = 13, 12, 11, 10$ are 8, 4, 4, 2 respectively. In each of these cases find the upper bounds given by the Singleton, Hamming and Plotkin bounds.

5

The power of linearity

5.1 The problem

In the previous chapter we (deliberately) forgot that error-correcting codes were developed as a solution to a practical problem, and now we return to the practicalities. Honesty demands that I admit to doing this only to motivate the next bit of nice mathematics, but for readers interested in the hardware design of encoders and decoders there are many good books with more emphasis on these matters. Two such texts are [11] and [12].

Suppose you have a code C which is sufficiently 'good', perhaps in the sense of the previous chapter, that you are tempted to use it to send and correct real messages. These messages could be English words, lines of a bank balance, directions for a robot, etc. Your first problem is encoding : how do you associate each of your possible messages with a codeword of C? To see that this is indeed a real problem just consider the fact that there could be millions of messages. You could just set up a 'look-up table' giving a list of all messages $m_1, m_2, \cdots$ with their corresponding codewords $c_1, c_2, \cdots$ just like an English–French dictionary: simple in principle but useless in practice because your encoder has to search the list every time you want to send a message, which is a hopelessly time-consuming task. Real dictionaries cut down the search time dramatically by a cunningly structured method of listing, the familiar alphabetic order. The Hamming code of Chapter 2 does much better by not requiring a look-up table at all: there was a simple, easily implementable algorithm for very rapidly translating each 4-bit message into its 7-bit codeword. The special mathematical structure by which the Hamming code solves both the space and time problems associated with look-up tables is the topic of this chapter.

For many nice codes the encoding problem is still unsolved. For example, several practical communications problems could be solved by using codes in which every codeword has the same weight, if only an efficient encoding algorithm could be invented.

There are similar problems at the other end of the channel. Assuming as usual that we use nearest neighbour decoding to correct transmission errors, for an arbitrary code we have to search the code to find the codeword(s) closest to the received word. Again the Hamming decoding process described in Chapter 2 reduces this search to a triviality. So what is so special about the Hamming code?

5.2 Linear codes – their fundamental properties

Most practical error-correcting codes in use today, including the Hamming codes, are examples of what are called *linear* codes. The main aim of this chapter is to explain what this means, and what consequences of linearity make such codes 'good'. This is the only part of the book which makes use of any assumed background knowledge – linear algebra, in particular the meaning and manipulation of vectors and matrices, and the elementary properties of a vector space. Reminders of the basic facts are given in section 5.3, but generally no proofs. If your memory of this has faded you should refer back to a text like [13] or to your favourite of the many linear algebra texts available.

One requirement of a linear code is that its alphabet symbols are the elements of a field. We shall use the fields Z_p, where p is prime, and call upon the material of Chapter 3. Before actually saying what a linear code is, here are just some of their advantages over arbitrary codes.

1. Evaluation of $d(C)$ is much easier.
2. Encoding is fast and requires little storage.
3. It is much easier to determine which errors are correctable/detectable.
4. The probability of correct decoding is much easier to calculate.
5. Very slick decoding techniques exist for linear codes.

Since words are n-strings of members of Z_p it is possible to define the 'sum' of two words and the product of a member of Z_p with a word. This is done as follows:

Definition 5.1

(i) $a_1a_2\cdots a_n + b_1b_2\cdots b_n = a_1+b_1,\ a_2+b_2\cdots a_n+b_n$

(ii) If $\alpha \in Z_p, \alpha(a_1a_2\cdots a_n) = \alpha a_1,\ \alpha a_2\ \cdots\ \alpha a_n$

Notice that if we think of the word $a_1a_2\cdots a_n$ as a vector with components $a_1, a_2, \cdots, a_n$, and α as a scalar, then these definitions are exactly the same as those for the vector sum and the scalar product. The only difference between this and the vector ideas with which you are probably more familiar is that both the scalars and the components of the vectors are members of Z_p rather than real or complex numbers, and of course all the arithmetic is modulo p. [Ex 1]

Now let C be a code of length n over Z_p.

Definition 5.2 C is a linear code over Z_p if for all $\mathbf{c}, \mathbf{c}' \in C$ and all $\alpha \in Z_p$

(i) $\mathbf{c} + \mathbf{c}' \in C$

and

(ii) $\alpha \mathbf{c} \in C$. [Ex 2–5]

Notice that Z_p^n is a vector space over Z_p so the definition of C as a linear code just amounts to saying C is a vector subspace of Z_p^n.

We now prove three easy consequences of linearity which are aesthetically pleasing and practically useful.

Theorem 5.1 All linear codes must contain the zero word.

Proof. Simply put $\alpha = 0$ in condition (ii). □

The main use of this is negative : if presented with a code for which $\mathbf{0} \notin C$ you know C cannot be linear!

Theorem 5.2 If C is linear $d(C)$ is the smallest weight of all the non-zero codewords.

Proof. Let $d(C) = d$ and let w be the smallest non-zero weight. Choose a pair of codewords $\mathbf{c}_1, \mathbf{c}_2$ with $d(\mathbf{c}_1, \mathbf{c}_2) = d$. (Note that $\mathbf{c}_1 \neq \mathbf{c}_2$ so $\mathbf{c}_1 - \mathbf{c}_2 \neq \mathbf{0}$) Then

$$d = d(\mathbf{c}_1, \mathbf{c}_2) = w(\mathbf{c}_1 - \mathbf{c}_2) \geq w \qquad (a)$$

Now choose a codeword $\mathbf{c}$ of weight w (Note $w \neq 0$) Then

$$w = w(\mathbf{c}) = w(\mathbf{c} - \mathbf{0}) = d(\mathbf{c}, \mathbf{0}) \geq d \qquad (b)$$

From (a) and (b) $d = w$ □

For an arbitrary code, finding the minimum distance involves examining every *pair* of codewords, but if the code is known to be linear, by the theorem above you only need to look at the weights of the individual codewords.

Definition 5.3 If C is any code over Z_p, $\mathbf{c}$ is a transmitted codeword, and $\mathbf{r}$ is the received word, then $\mathbf{e} = \mathbf{r} - \mathbf{c}$ is called the *error pattern* of this transmission.

Rewriting the equation in the above definition as $\mathbf{r} = \mathbf{c} + \mathbf{e}$, we can think of $\mathbf{e}$ as the 'noise' which acts on $\mathbf{c}$ to convert it to $\mathbf{r}$.

Theorem 5.3 For a linear code using nearest neighbour decoding, whether or not a received word $\mathbf{r}$ is uniquely correctly decodable depends only on $\mathbf{e}$, not $\mathbf{r}$.

Proof. In this proof we insert a couple of diagrams which treat words (vectors in Hamming space) as if they were geometric displacement or position vectors. Of course they are not, and this device is merely a visual aid for the proof, not a mathematically necessary part of it.

The proof is by contradiction : we suppose that there are codewords $\boldsymbol{c}, \boldsymbol{c}'$ of a linear code, and an error pattern $\boldsymbol{e}$ such that $\boldsymbol{r} = \boldsymbol{c} + \boldsymbol{e}$ is correctly decoded but $\boldsymbol{r}' = \boldsymbol{c}' + \boldsymbol{e}$ is not. The fact that $\boldsymbol{r}'$ is incorrectly decoded means that there is some codeword $\boldsymbol{c}''$ at least as close to $\boldsymbol{r}'$ as $\boldsymbol{c}'$ is. Let $\boldsymbol{e}'$ be the error pattern associated with transmitting $\boldsymbol{c}''$ and receiving $\boldsymbol{r}'$. We have

$$\boldsymbol{e}' = \boldsymbol{r}' - \boldsymbol{c}'' \text{and} d(\boldsymbol{c}'', \boldsymbol{r}') \leq d(\boldsymbol{c}', \boldsymbol{r}'),$$

so the second of these relations can be written $w(\boldsymbol{e}') \leq w(\boldsymbol{e})$. Finally, define $\boldsymbol{u}$ as the word $\boldsymbol{r} - \boldsymbol{e}'$. All this is illustrated on the diagram.

Now

$$\begin{aligned} \boldsymbol{u} &= \boldsymbol{r} - \boldsymbol{e}' = \boldsymbol{c} + \boldsymbol{e} - \boldsymbol{e}' = \boldsymbol{c} + (\boldsymbol{r}' - \boldsymbol{c}') - (\boldsymbol{r}' - \boldsymbol{c}'') \\ &= \boldsymbol{c} - \boldsymbol{c}' + \boldsymbol{c}'' \end{aligned}$$

which is in C because $\boldsymbol{c}$, $\boldsymbol{c}'$ and $\boldsymbol{c}''$ and all in $C$are C is linear. But $d(\boldsymbol{u}, \boldsymbol{r}) = w(\boldsymbol{e}') \leq w(\boldsymbol{e}) = d(\boldsymbol{c}, \boldsymbol{r})$. So $\boldsymbol{r}$ is not uniquely decoded to $\boldsymbol{c}$ and we have our contradiction. □

To conclude this section we have a nice symmetry property of linear codes, but first an exercise to set the scene. [Ex 6]

The result of Exercise 6 can be generalized to all linear codes:

Theorem 5.4 Let C be a linear code over Z_p. In each position C either has 0 in every codeword, or each of the p symbols occur equally often.

Proof. We fix attention on the first position since the argument is identical for all positions, and suppose 0 is not the first symbol of every codeword. Let C be $\{\mathbf{c}_1, \mathbf{c}_2, \cdots, \mathbf{c}_M\}$ and let $\mathbf{c}_1$ have a $(\neq 0)$ as its first symbol.

By the methods of Chapter 3 we see that $0a, 1a, 2a, \cdots, (p-1)a$ are just the numbers $0, 1, 2, \cdots, p-1$ in some order (with arithmetic being modulo p), and by linearity, $0\mathbf{c}_1, 1\mathbf{c}_1, 2\mathbf{c}_1, \cdots, (p-1)\mathbf{c}_1$ are all codewords and their first symbols are $0a, 1a, 2a, \cdots, (p-1)a$ respectively. Hence C contains at least one word beginning with each of the p symbols.

Now let b be any one of the non-zero symbols, and let $\mathbf{c}$ be a codeword beginning with b. Then $\{\mathbf{c}+\mathbf{c}_1, \mathbf{c}+\mathbf{c}_2, \cdots, \mathbf{c}+\mathbf{c}_M\}$ is a set of distinct words, all in C by linearity, so this set is in fact C. Also $\mathbf{c} - \mathbf{c}_i$ begins with b if and only if $\mathbf{c}_i$ begins with 0. Hence the number of codewords starting with b is the same as the number starting with 0. b was arbitrary so the theorem is proved. □

5.3 Linear algebra reminders

The following is a list of definitions and results from linear algebra, but phrased, where necessary, in the language of codes as explained in section 5.2. We regard Z_p^n as a vector space over Z_p and a linear code with Z_p as its alphabet as a vector subspace of Z_p^n.

1. A *linear combination* of all words $\boldsymbol{x}_1, \boldsymbol{x}_2, \cdots, \boldsymbol{x}_t$ is a word of the form $\sum_{i=1}^{t} \lambda_i\, \boldsymbol{x}_i$ where each $\lambda_i \in Z_p$.

   The λ_i will be called the coefficients of the linear combination.

2. The *span* of the words $\boldsymbol{x}_1, \boldsymbol{x}_2, \cdots, \boldsymbol{x}_t$, written $\langle \boldsymbol{x}_1, \boldsymbol{x}_2, \cdots, \boldsymbol{x}_t \rangle$ is the set of all linear combinations of these words.

   If S and T are sets of words for which $\langle S \rangle \supseteq TS$ is said to span T. [Ex 7]

3. If $\boldsymbol{x} = x_1 x_2 \cdots x_n$ and $\boldsymbol{y} = y_1\, y_2 \cdots y_n$ are words of Z_p^n the *dot product* or *inner product* of $\boldsymbol{x}$ and $\boldsymbol{y}$, written $\boldsymbol{x} \cdot \boldsymbol{y}$, is the number

$$x_1 y_1 + x_2 y_2 + \cdots + x_n y_n \text{mod } p.$$

4. $\boldsymbol{x}$ and $\boldsymbol{y}$ are called orthogonal words if $\boldsymbol{x} \cdot \boldsymbol{y} = 0$. [Ex 8, 9]

5. For any $S \subseteq Z_p^n S^{\perp}$ (pronounced S perp) is the set of all words which are orthogonal to every word of S [Ex 10, 11]

6. If C is a linear code then $C^{\perp}$ is called the *dual code of* C, and by Exercise 10, $C^{\perp}$ is also linear.

7. A set of words S is called a *linearly independent set* if $\mathbf{0}$ can *only* be expressed as a linear combination of them by taking every coefficient to be 0. Otherwise S is a *dependent set.* [Ex 12, 13]

8. If C is a linear code, then any subset of C which is linearly independent and which spans C is called a *basis of* C. Equivalent definitions of a basis of C are:

   (a) any $B \subseteq C$ where B is independent and has the property that putting any other member of C into B makes it a dependent set;

   (b) any $B \subseteq C$ which has the property that B spans C but no proper subset of B spans C.

   (a) and (b) are sometimes expressed as 'B is a maximal independent subset of C' and 'B is a minimal spanning set of C' respectively.

9. Bases have the following properties:

   (a) Given any spanning set S for C there is a subset of S which is a basis of C;

   (b) Given any linearly independent subset I of C, there is a set J such that $I \cup J$ is a basis for C;

   (c) All bases of C have the same number of words. This number is called the *dimension* of C. [Ex 14]

10. For all primes p, Z_p^n has dimension n because the unit words $\boldsymbol{w}_1, \boldsymbol{w}_2, \cdots,$ $\boldsymbol{w}_n$ (where $\boldsymbol{w}_i$ has 1 as its ith component and 0 for all others) clearly span Z_p^n and are linearly independent.

    It follows from (9) and (10) that any set of more than n words in Z_p^n must be a dependent set, and no set of fewer than n words of Z_p^n can span Z_p^n.

11. Let C_1, C_2 be linear codes of length n over Z_p, with dimensions k_1, k_2 respectively.

    (a) If $C_1 \subseteq C_2$ then $k_1 \leq k_2$ and

    (b) if $C_1 \subseteq C_2$ and $k_1 = k_2$, then $C_1 = C_2$. [Ex 15–17]

    NOTATION: The dimension of C, a linear code over Z_p, is often written as dim(C). If dim (C) $= k$, then M, the number of codewords of C, depends (from Exercise 16) only on k and p. It is customary when dealing with *linear* codes to describe a code as a $p-ary[n, k, d]$ code rather than a $p - ary(n, M, d)$ code – or just an $[n, k]$ code if p is clear from the context and d is either not known or not specified.

12. The rank-nullity theorem.
    If A is any $k \times n$ matrix (with entries in Z_p), the *image space* of A, $Im(A)$ is the span of the rows of A, and the *null space* of A, Null (A) is the set of all words $\boldsymbol{w}$ for which $\boldsymbol{w}A^T = \mathbf{0}$, ($A^T$ denotes the transpose of A). The *rank* of A is dim ($Im(A)$) and the *nullity* of A is dim ($Null(A)$). The *rank-nullity theorem* states that

$$dim(Null(A)) + dim(Im(A)) = n.$$

13. If X, Y are the sets of rows and columns, respectively, of any matrix, then dim $\langle X \rangle =$ dim $\langle Y \rangle$.

5.4 The generator matrix

Remember that the whole point of encoding messages is to introduce redundancy with a view to doing error detection or correction. How then are messages encoded when using a linear code? We normally take the set of all messages (the *message space*) to be all words of a given length $n - k$, and encode them into codewords of length n. This is done using a linear code of length n and dimension k. (So $k \leq n$ by (10) and (11) of section 5.3). Then we take a matrix G whose rows are the codewords of any basis of C, say $\mathbf{c}_1, \mathbf{c}_2, \cdots \mathbf{c}_k$, and define for each message $\boldsymbol{m}$ the corresponding codeword $\mathbf{c}$ by $\mathbf{c} = \boldsymbol{m}\ G$.

For example $\mathbf{c}_1 = 12043$, $\mathbf{c}_2 = 23104$, $\mathbf{c}_3 = 40211$, is a basis for a 3 dimensional code over Z_5 (we leave you to check that these words are independent), so the message $\boldsymbol{m} = 123$ is encoded as

$$\boldsymbol{m}G = (1\ \ 2\ \ 3)\begin{pmatrix} 1 & 2 & 0 & 4 & 3 \\ 2 & 3 & 1 & 0 & 4 \\ 4 & 0 & 2 & 1 & 1 \end{pmatrix} = 23324.$$

So C is just the span of the rows of $G, Im(G)$. [Ex 18]

Particularly convenient linear codes are those which have a $k \times n$ generator matrix in which the first k columns just make up the $k \times\ k$ unit matrix I_k, because in this case the message coincides with the first k symbols of its codeword. This is a time-saving feature because the vast majority of words are received with no errors so for these words all the receiver needs to do with the received codeword is read off its first k symbols. Recall that the 7-bit and 8-bit Hamming codes of Chapter 2 had just this property, and we shall see later that these codes are in fact linear, and the encoding which we did then with Venn diagrams can be done with a generator matrix.

To return to our previous example, C is a [5, 3] code so $d(C) \leq 5$. This can be strengthened by using Theorem 5.2 and the fact that the rows of G are codewords and these have weight 4, so $d(C) \leq 4$. Short of listing all 125 codewords it is not easy to see what is the minimum non-zero weight of the codewords, so an important question is : how can a linear code be deliberately designed to have a specified minimum distance? See later this chapter and the next for an answer. [Ex 19]

5.5 Cosets and the Slepian array

The point of this section is to show a way of implementing nearest neighbour decoding for linear codes. The following example illustrates all the significant features of the method.

Take C to be the [3, 2] code over Z_3 generated by

$$G = \begin{bmatrix} 2 & 0 & 1 \\ 1 & 1 & 2 \end{bmatrix}.$$

So C is the set of nine words, $\{\lambda(2\ \ 0\ \ 1)+\mu(1\ \ 1\ \ 2) : 0\ \leq \lambda \leq 2, 0 \leq \mu \leq 2\}$, which we write as a row of codewords in arbitrary order,

201 112, 000, 102, 221, 010, 122, 211, 020.

Then pick any word not in C (I chose 222) and write down as a second row the words formed by adding 222 to each of the words in the first row. This set of words is called the *coset* $C + 222$. The first two rows still do not include all words of Z_3^3 so pick one of the missing ones (I chose 200) and add this to all the codewords to make the third row $C + 200$. The result is:

$$
\begin{array}{rcccccccccc}
C & = & 201 & 112 & 000 & 102 & 221 & 010 & 122 & 211 & 020 \\
C + 222 & = & 120 & 001 & 222 & 021 & 110 & 202 & 011 & 100 & 212 \\
C + 200 & = & 101 & 012 & 200 & 002 & 121 & 210 & 022 & 111 & 220
\end{array}
$$

We stop here because our array contains every word of Z_3^3. Notice that the rows are pairwise disjoint and no row contains any repeated words.

[Ex 20, 21]

Exercise 21 establishes the principal properties of the cosets of a linear code. It can be summed up by saying that the distinct cosets partition Z_p^n into p^{n-k} cosets each of size p^k.

To implement the advertized decoding method we first make two further restrictions on the array of C and its cosets. One is that the first row (of codewords of C) must start with the zero codeword. The other is that the first word of each other row must be chosen from the words of smallest weight not already included in any previous row. An array satisfying these conditions for the same code C is shown below:

$$
\begin{array}{rcccccccccc}
C & = & 000 & 201 & 112 & 102 & 221 & 010 & 122 & 211 & 020 \\
C + 100 & = & 100 & 001 & 212 & 202 & 021 & 110 & 222 & 011 & 120 \\
C + 200 & = & 200 & 101 & 012 & 002 & 121 & 210 & 022 & 111 & 220
\end{array}
$$

This is called a Slepian (or standard) array. The words chosen to go in the first column are called the coset leaders. [Ex 22]

The decoding process has just two steps.

Slepian array decoding: 1. Locate the received word in the array.

2. Decode it as the codeword at the top of its column.

Why this is sensible is explained by the next theorem.

Theorem 5.5 Slepian array decoding is nearest neighbour decoding.

Proof. Let r be the received word, l its coset leader, and c the codeword at the top of r's column.

$$
\begin{array}{cccccccccccccccc}
\cdot & \cdot & \cdot & \cdot & \cdot & c' & \cdot & \cdot & \cdot & c & \cdot & \cdot & \cdot & \cdot & \cdot & \cdot \\
\cdot & \cdot & \cdot & \cdot & \cdot & \cdot & \cdot & \cdot & \cdot & \cdot & \cdot & \cdot & \cdot & \cdot & \cdot & \cdot \\
\cdot & \cdot & \cdot & \cdot & \cdot & \cdot & \cdot & \cdot & \cdot & \cdot & \cdot & \cdot & \cdot & \cdot & \cdot & \cdot \\
l & \cdot & \cdot & \cdot & \cdot & \cdot & \cdot & \cdot & r & \cdot & \cdot & \cdot & e & \cdot & \cdot & \cdot \\
\cdot & \cdot & \cdot & \cdot & \cdot & \cdot & \cdot & \cdot & \cdot & \cdot & \cdot & \cdot & \cdot & \cdot & \cdot & \cdot \\
\cdot & \cdot & \cdot & \cdot & \cdot & \cdot & \cdot & \cdot & \cdot & \cdot & \cdot & \cdot & \cdot & \cdot & \cdot & \cdot
\end{array}
$$

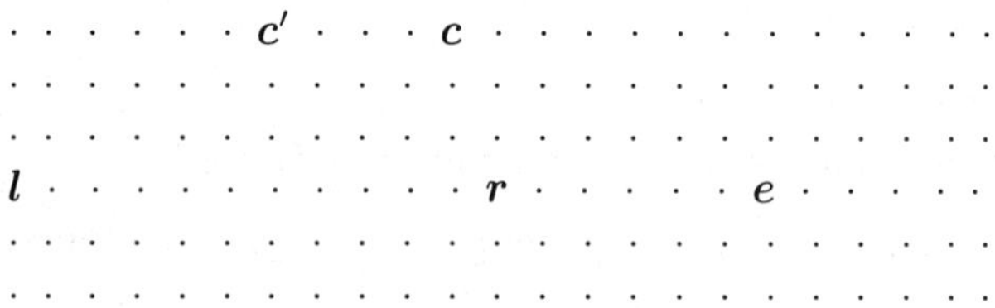

We have to show that $d(r, c) \le d(r, c')$ for all codewords c'. With the aim of deducing a contradiction suppose there is a codeword c' with

$$d(r, c) > d(r, c') \tag{1}$$

We have

$$r = c + l \qquad (2)$$

by the rule for constructing Slepian arrays. Let

$$r = c' + e \qquad (3)$$

From (2) and (3) we derive

$$e - l = (r - c') - (r - c) = c - c'$$

So $e - l \in C$ by linearity.

Hence, by Exercise 20(iii) e and l are in the same coset (same row of the array).

But using (2) and (3) and Exercise 4 in (1) we get $w(r - c') < w(r - c)$. That is, $w(e) < w(l)$ which contradicts the rules for the array since l is a coset leader and therefore has weight no bigger than that of any word in its row. □

We have seen previously that if a code has minimum distance d then nearest neighbour decoding, using the complete decoding scheme described in section 4.3 will correct all instances of $\leq [\frac{d-1}{2}]$ errors but may also correct some received words with a greater number of errors. Since Slepian array decoding is a version of nearest neighbour decoding we can ask which error patterns precisely will the method correctly decode?

Theorem 5.6 The error patterns correctly decoded by a Slepian array are the coset leaders.

Proof. Let c be sent, corrupted by a channel error e, so that $r = c + e$ is received. r is correctly decoded (to c) if and only if r and c are in the same column. That is, $r = c + l$ for some coset leader, l. i.e. $c + e = c + l$, so $e = l$ □

[Ex 23]

Now we have a problem : the choice of coset leaders is not necessarily unique because the set of words not covered by the first i rows could contain several words of the same (minimal) weight. So, from the previous theorem, which error patterns are corrected depends on the choice of coset leaders. The point of the next theorem is to show that this is not a serious problem.

Theorem 5.7 Let S, S' be different Slepian arrays for the linear code C. Let S_i be the set of cosets in S whose leaders have weight i, and let S'_i be the corresponding set of S'. Then for all i, $S_i = S'_i$.

Proof. Let $C + x$ be any coset in S_i. Then by the properties of cosets developed in Exercise 21, $C + x$ is also a coset of S'. Furthermore $C + x$ contains words of weight i but none of weight less then i, so $C + x$ must also be a coset of S'. Hence $S_i \subseteq S'_i$ and by a similar argument $S'_i \subseteq S_i$, So $S_i = S'_i$. □

This theorem ensures that, given any linear code, the *numbers* of error patterns of weights $0, 1, 2, \cdots$ which it can correct by Slepian array decoding does not depend on which Slepian array is used. [Ex 24–27]

5.6 The dual code and parity check matrix

We now return to the Hamming code of Chapter 2 to introduce an alternative method of specifying a linear code. Recall that the messages were all 16 4-bit binary words which were then encoded as 7-bit codewords. The three additional bits were fixed by requiring that the total number of ones in each of the three sets was even. Using modulo 2 addition and denoting a codeword by $x_1\ x_2\ \cdots\ x_7$ these conditions become:

$$\begin{array}{ccccccccccccccc} x_1 & & & + & x_3 & + & x_4 & + & x_5 & & & & & = & 0 \\ x_1 & + & x_2 & & & + & x_4 & & & + & x_6 & & & = & 0 \\ x_1 & + & x_2 & + & x_3 & & & & & & & + & x_7 & = & 0 \end{array}$$

So this Hamming code C can be specified as the set of all 7-bit strings which satisfy these equations. The equations can be written compactly in matrix form as

$$\mathbf{c}H^T = \mathbf{0}$$

where $\mathbf{c}$ is the codeword $(x_1 x_2 \cdots x_7)$, regarded as a row vector, H is the matrix

$$\begin{bmatrix} 1 & 0 & 1 & 1 & 1 & 0 & 0 \\ 1 & 1 & 0 & 1 & 0 & 1 & 0 \\ 1 & 1 & 1 & 0 & 0 & 0 & 1 \end{bmatrix},$$

$\mathbf{0}$ is the zero column vector

$$\begin{pmatrix} 0 \\ 0 \\ 0 \end{pmatrix}$$

and H^T is the transpose of H. Notice that the left hand sides of the three equations are just the dot products of $\mathbf{c}$ with the rows of H, so another description of C is that it is just $S^{\perp}$ where S is $\{1011100, 1101010, 1110001\}$, so we know from Exercise 11 that C is linear. [Ex 28]

Definition 5.4 H is called a parity check matrix for a linear code C if

(i) its rows are independent,

(ii) C is the set of all words satisfying $\mathbf{c}H^T = \mathbf{0}$. (That is, C is the null space of H.)

We now show that every linear code C has a parity check matrix, but first we need another property of $C^{\perp}$, whose proof you will find easier if you have met the idea of a singular matrix. If not, ignore the proof but make sure you understand the results.

Theorem 5.8 If C is an $[n,k]$ code over Z_p, then $C^{\perp}$ is an $[n,n-k]$ code.

Proof. We already know that $C^{\perp}$ is linear so only the fact that its dimension is $n-k$ remains to be proved.

Choose a basis $\{\boldsymbol{v}_1,\cdots,\boldsymbol{v}_n\}$ for Z_p^n in which $\{\boldsymbol{v}_1,\cdots,\boldsymbol{v}_k\}$ is a basis for C. (See 9(i) of section 5.3) Let B be the $n\times n$ matrix whose rows are the $\boldsymbol{v}_i$. Then the matrix G consisting of the first k rows of B is a generator matrix for C. B is non-singular so it has an inverse B^{-1}. Let $\boldsymbol{w}_1,\boldsymbol{w}_2,\cdots,\boldsymbol{w}_n$ be the rows of $(B^{-1})^T$. $(B^{-1})^T$ is non-singular so $\{\boldsymbol{w}_1,\cdots,\boldsymbol{w}_n\}$ is another basis of Z_p^n. $C^{\perp}$ is actually the span of the last $n-k$ of these rows, as we now demonstrate, so that these words make a basis of $C^{\perp}$.

We first prove that $\boldsymbol{w}_{k+1},\cdots,\boldsymbol{w}_n$ are all in $C^{\perp}$ by showing that each of them is orthogonal to every word of C. Let $\boldsymbol{v}\in C$ so that it can be expressed as $\sum_{i=1}^{k}\alpha_i\boldsymbol{v}_i$. Then for $k+1\le j\le n$,

$$\begin{aligned}
\boldsymbol{v}.\boldsymbol{w}_j &= \left(\sum_{i=1}^{k}\alpha_i\boldsymbol{v}_i\right)\cdot\boldsymbol{w}_j=\sum_{i=1}^{k}\alpha_i(\boldsymbol{v}_i\cdot\boldsymbol{w}_j)\\
&= \sum_{i=1}^{k}\alpha_i\left((\text{row } i \text{ of } B)\cdot(\text{row } \ j \text{ of } \ (B^{-1})^T)\right)\\
&= \sum_{i=1}^{k}\alpha_i\left((\text{row } i \text{ of } B)\cdot(\text{column } \ j \text{ of } \ (B^{-1}))\right)\\
&= \sum_{i=1}^{k}\alpha_i\left(i,j \text{ entry of } BB^{-1}\right)\\
&= 0
\end{aligned}$$

since $BB^{-1}=I$, whose i, j entry is 0 because $j>k\ge i$. By a similar argument we now establish that all words of $C^{\perp}$ can be written as a linear combination of $\boldsymbol{w}_{k+1},\cdots,\boldsymbol{w}_n$.

Let $\boldsymbol{w}\in C^{\perp}$, so write it as

$$\boldsymbol{w}=\sum_{i=1}^{n}\beta_i\boldsymbol{w}_i \qquad (1)$$

We aim to show $\beta_1=\beta_2=\cdots\cdots=\beta_k=0$. Now $\boldsymbol{v}_1,\cdots\boldsymbol{v}_k$ are all in C, so $\boldsymbol{w}\cdot\boldsymbol{v}_j=0\text{for}j=1,2,\cdots,k$. That is, for

$$\begin{aligned}
1\le\ j\le k,0 &= \sum_{i=1}^{n}(\beta_i\boldsymbol{w}_i)\cdot\boldsymbol{v}_j=\sum_{i=1}^{n}\beta_i(\boldsymbol{w}_i\cdot\boldsymbol{v}_j)\\
&= \sum_{i=1}^{n}\beta_i(\text{row } i \text{ of } (B^{-1})^T)\cdot\ (\text{row } j \text{ of } B)
\end{aligned}$$

$$= \sum_{i=1}^{n} \beta_i \,(j, i \text{ entry of } BB^{-1})$$
$$= \beta_j$$

since the j, i entry of BB^{-1} is 1 when $i = j$ and 0 otherwise. So (1) becomes

$$\boldsymbol{w} = \sum_{i=k+1}^{n} \beta_i \boldsymbol{w}_i.$$

We have shown that the last $n - k$ rows of $(B^{-1})^T$ are independent members of $C^{\perp}$, and each member of $C^{\perp}$ is a linear combination of them, so $\dim(C^{\perp}) = n - k$. □

Now with the aid of the important rank-nullity theorem (item 12 of section 5.3) we find a neat connection between the parity check and generator matrices.

Theorem 5.9 H is a parity check (p.c.) matrix for the $[n, k]$ code C if and only if it is a generator matrix for $C^{\perp}$.

Proof.

(i) H is a parity check matrix for C
$\Rightarrow C = \text{null}(H)$
$\Rightarrow k = n - \dim\,(Im(H))$ by the rank-nullity theorem
$\Rightarrow \dim\,(Im(H)) = n - k$
$\Rightarrow \dim\,(Im(H)) = \dim(C^{\perp})$ by the previous theorem.
But $Im(H) \subseteq C^{\perp}$, so $Im(H) = C^{\perp}$ by (11) of section 5.3. That is, H is a generator matrix for $C^{\perp}$.

(ii) H is a generator matrix for $C^{\perp}$
$\Rightarrow$ rows of H are independent and $C^{\perp} = Im(H)$
$\Rightarrow \boldsymbol{c}$.(any linear combination of rows of H) $= 0$ for all $\boldsymbol{c} \in C$.
$\Rightarrow C \subseteq \text{Null}\ (H)$
$\Rightarrow C = \text{Null}\ (H)$ by (11) of section 5.3 since

$$\dim(C) = k = n - (n-k) = n - \dim C^{\perp}$$
$$= n - \dim(ImH) = \dim(\text{null}\ (H))$$

That is, H is a parity check matrix for C. □

From this result it follows that every linear code C has a parity check matrix – simply use any generator matrix of $C^{\perp}$. [Ex 29–32]

5.7 Syndrome decoding

The method of decoding by Slepian array is neat conceptually but suffers from two major drawbacks when used with large codes. The first is the *space problem* : if C has lots of codewords it takes up lots of memory space to store it in a computer, and the array is p^{n-k} times bigger than C! The second is the *time problem*: even if the whole array could be stored the first step of the decoding process is to search the array to find the received word, and this could be very time consuming. The method of syndrome decoding solves both of these problems whilst retaining the spirit of array decoding.

Definition 5.5 Let H be a parity check matrix for a linear code, and let $\boldsymbol{v}$ be any word. The *syndrome* of $\boldsymbol{v}$, syn $(\boldsymbol{v}) = \boldsymbol{v}H^T$.

By our previous results syn $(\boldsymbol{v}) = \mathbf{0}$ if and only if $\boldsymbol{v}$ is a codeword. Furthermore,

Theorem 5.10 $\boldsymbol{u}, \boldsymbol{v}$ belong to the same coset if and only if syn $(\boldsymbol{u}) =$ syn$(\boldsymbol{v})$.

Proof. $\boldsymbol{u}, \boldsymbol{v} \in$ same coset $\Leftrightarrow \boldsymbol{u} - \boldsymbol{v} \in C$ by Exercise 21

$\Leftrightarrow (\boldsymbol{u} - \boldsymbol{v})H^T = \mathbf{0}$ by definition of p.c. matrix

$\Leftrightarrow \boldsymbol{u}H^T - \boldsymbol{v}H^T$

$\Leftrightarrow$ syn$(\boldsymbol{u}) =$ syn$(\boldsymbol{v})$. □

This means that each row of a Slepian array (each coset) consists of all those words which have the same syndrome (that of the coset leader).

The next theorem connects the columns of a p.c. matrix with the syndrome of any received word.

Theorem 5.11 Let C be an $[n, k]$ code and H be any of its p.c. matrices. If $\boldsymbol{e} = e_1 \cdots e_n$ is the error pattern associated with the received word $\boldsymbol{r}$, then syn $(\boldsymbol{r}) = \left(\sum_{i=1}^{n} e_i \boldsymbol{h}_i\right)^T$ where $\boldsymbol{h}_i$ is the ith column of H.

Proof.
$$\text{syn}(\boldsymbol{r}) = \text{syn}(\boldsymbol{e}) = \boldsymbol{e}H^T = (e_1 \cdots e_n) \begin{bmatrix} h_{11} & \cdot & \cdot & \cdot & \cdot & h_{n-k1} \\ \cdot & & & & & \\ \cdot & & & & & \\ \cdot & & & & & \\ h_{1n} & \cdot & \cdot & \cdot & \cdot & h_{n-kn} \end{bmatrix}$$

$$= (e_1 h_{11} + \cdots + e_n h_{1n}, e_1 h_{21} + \cdots + e_n h_{2n}, \cdots, e_1 h_{n-k1} + \cdots + e_n h_{n-kn})$$

$$= e_1 (h_{11}, \cdots h_{n-k1}) + \cdots + e_n (h_{1n}, \cdots, h_{n-kn})$$

$$= e_1 \boldsymbol{h}_1^T + \cdots + e_n \boldsymbol{h}_n^T = \left(\sum e_i \boldsymbol{h}_i\right)^T$$ □

A special case of this result is that for binary codes syn$(\boldsymbol{r})$ is the transpose of the sum of those columns of H corresponding to the positions where the errors occur.

The space problem associated with Slepian array decoding was that the whole array is needed throughout the use of the code, so that a large chunk of computer memory is permanently occupied. In syndrome decoding the array is calculated prior to using the code, and then all except the first column (the coset leaders) can be thrown away. All that needs to be stored permanently are the list of coset leaders, their associated syndromes and the p.c. matrix.

The syndrome decoding steps are then very simple:

1. Calculate the syndrome $\boldsymbol{r}H^T$ of the received word $\boldsymbol{r}$.

2. Scan the stored list to find the coset leader $\boldsymbol{e}$ with the same syndrome.

3. Decode to the codeword $\boldsymbol{c} = \boldsymbol{r} - \boldsymbol{e}$.

This is clearly equivalent to array decoding because $\boldsymbol{e}$ is the leader of the row in which $\boldsymbol{r}$ would have appeared had the whole array been stored, and $\boldsymbol{r} - \boldsymbol{e}$ would be the codeword at the top of the column containing $\boldsymbol{r}$. [Ex 33]

The results of Exercise 33 are both consequences of the next general result which characterizes $d(C)$ in terms of the columns of H.

Theorem 5.12 The minimum distance d of a linear code C is the size of the smallest dependent set of columns of H.

Proof. Let $\boldsymbol{c} = c_1 \cdots c_n$ be a codeword of weight d (which exists by Theorem 5.2.) Now $\boldsymbol{c}H^T = \boldsymbol{0}$, which can be expanded as the equation $c_1\boldsymbol{h}_1 + \cdots + c_n\boldsymbol{h}_n = \boldsymbol{0}$, $\boldsymbol{h}_i$ being the ith column of H, and the equation has only d non-zero terms. The corresponding set of columns $\{\boldsymbol{h}_i : c_i \neq 0\}$ is therefore dependent.

To show that there is no smaller dependent set, consider any set of t dependent columns $\{\boldsymbol{h}_{\alpha_1}, \cdots, \boldsymbol{h}_{\alpha_t}\}$. There are constants $k_{\alpha_1}, \cdots, k_{\alpha_t}$, not all zero, such that $k_{\alpha_1}\boldsymbol{h}_{\alpha_1} + \cdots + k_{\alpha_t}\boldsymbol{h}_{\alpha_t} = \boldsymbol{0}$, so that the word $\boldsymbol{x}$ with k_{α_i} as the α_i th component and the rest zero is a codeword of weight $\leq t$. Hence $t \geq d$ by Theorem 5.2 again, and the result is proved. □

In the next chapter we shall use this result to design an important class of codes with minimum distance 3.

Sometimes it is possible to construct the coset leader list without having to construct the whole Slepian array. The following example will illustrate the arguments used.

Let C be a ternary $[7, 3]$ code with

$$H = \begin{bmatrix} 1 & 2 & 0 & 1 & 0 & 0 & 0 \\ 1 & 1 & 1 & 0 & 1 & 0 & 0 \\ 0 & 0 & 1 & 0 & 0 & 1 & 0 \\ 2 & 2 & 2 & 0 & 0 & 0 & 1 \end{bmatrix}$$

It is easy to check that no pair of columns is dependent, but the three columns 1, 2 and 4 are dependent since

$$2\begin{pmatrix}1\\1\\0\\2\end{pmatrix}+1\begin{pmatrix}2\\1\\0\\2\end{pmatrix}+2\begin{pmatrix}1\\0\\0\\0\end{pmatrix}=\begin{pmatrix}0\\0\\0\\0\end{pmatrix}.$$

So by the previous theorem $d(C) = 3$ so C is 1-error-correcting, so **0** and all words of weight 1 are coset leaders. This accounts for 15 cosets. $dim(C) = 3$ so there are $3^3 = 27$ codewords, and Z_3^7 has 3^7 words, so there must be $\frac{3^7}{3^3} = 81$ cosets. To find the remaining 66 coset leaders we could systematically work through the words of weight 2. Each time we find one which has a syndrome not included in the list so far, add this as a new coset leader. Notice that there are $\binom{7}{2} \times 2^2 = 84$ words of weight 2, so not all of them are coset leaders. (Another reason for this is that C is not 2-error correcting.) Furthermore, it may be that some cosets contain lots of words of weight ≥ 2 so that all words of weight ≤ 2 yield fewer than 81 distinct syndromes. In this case the search has to be widened to words of weight 3, and so on. We leave you to get a full list of coset leaders and possibly find a better method of doing the search. [Ex 34]

It is time we did an example of syndrome decoding. Taking the code C' of Exercise 22 and its solution we see that all non-zero codewords have weight 3 so $d(C') = 3$. Hence C' is one-error correcting, which implies that each of the 8 words of weight 1 is a coset leader. There are 9 leaders in all, **0** has to be one, so this accounts for all the coset leaders.

Now we need the syndromes of the coset leaders, and to do this we need a p.c. matrix. C' is a $[4,2]$ code, so by Theorem 5.8 $C'^{\perp}$ is also $[4,2]$. So the two rows of a p.c. matrix for C can be any two independent rows which are orthogonal to each row of G. Hence words $x_1x_2x_3x_4$ of $C'^{\perp}$ have to satisfy

$$\left.\begin{array}{lrcrcrcrcl} & & & x_2 & + & 2x_3 & + & x_4 & = & 0 \\ \text{and} & x_1 & + & 2x_2 & + & 2x_3 & & & = & 0 \end{array}\right\}\text{modulo } 3.$$

Two independent solutions are 1102 and 2110 so we may take H to be

$$\begin{bmatrix} 1 & 1 & 0 & 2 \\ 2 & 1 & 1 & 0 \end{bmatrix},$$

and the list of coset leaders and syndromes are:

coset leader:	0000	1000	2000	0100	0200	0010	0020	0001	0002
syndrome:	00	12	21	11	22	01	02	20	10

Suppose 2221 is received. Its syndrome is 02 and the corresponding coset leader is 0020, so 2221 is decoded as 2221 – 0020 = 2201, and if there is indeed only one error in the received word, this decoding will be correct.

There is an important point to make about the example above. The example was unrealistically easy because the number of cosets matched exactly the number of weight 0 and 1 words, and because $d = 3$ we knew that all such words would be coset leaders and that there were no more. Secondly much of the work was in finding H from G – not too bad for this example but very hard work for big codes. We develop on algorithm to alleviate this.

Definition 5.6 A generator matrix is in **standard form** if it has the form $[I|A]$ where I is an identity matrix.

Theorem 5.13 If the $[n, k]$ code C has standard generator matrix $G = [I|A]$, then a p.c. matrix for C is $[-A^T|I]$. For example, if

$$G = \begin{bmatrix} 1 & 0 & 0 & 2 & 1 & 4 & 0 \\ 0 & 1 & 0 & 3 & 0 & 1 & 2 \\ 0 & 0 & 1 & 1 & 2 & 3 & 2 \end{bmatrix}$$

for a $[7, 3]$ code over Z_5, then

$$H = \begin{bmatrix} -2 & -3 & -1 & 1 & 0 & 0 & 0 \\ -1 & -0 & -2 & 0 & 1 & 0 & 0 \\ -4 & -1 & -3 & 0 & 0 & 1 & 0 \\ -0 & -2 & -2 & 0 & 0 & 0 & 1 \end{bmatrix} = \begin{bmatrix} 3 & 2 & 4 & 1 & 0 & 0 & 0 \\ 4 & 0 & 3 & 0 & 1 & 0 & 0 \\ 1 & 4 & 2 & 0 & 0 & 1 & 0 \\ 0 & 3 & 3 & 0 & 0 & 0 & 1 \end{bmatrix}$$

is a p.c. matrix. You should check this by verifying that each row of H is orthogonal to each row of G (and hence that each word in the span of the rows of H is orthogonal to each word in the span of the rows of G). If you do this using the form for H with the negative entries, you will see why it works in general. □

For reasons which will be apparent in the next section it is often useful to be able to get from G to H by a quick method such as that explained in the previous theorem in those cases where G is not exactly in standard form. The method is daunting to write out in full generality so we shall be content with a representative example.

Let G have the same columns as in the previous example, but in a different order:

$$G = \begin{bmatrix} 2 & 1 & 0 & 4 & 0 & 0 & 1 \\ 3 & 0 & 2 & 1 & 0 & 1 & 0 \\ 1 & 0 & 2 & 3 & 1 & 0 & 2 \end{bmatrix}$$

Notice the 'unit columns' are columns 2, 5 and 6. The first step in constructing H is to fill in its columns 2, 5 and 6. Look at column 2 of G. This has its 1 in position 1 so fill column 2 of H with the remaining entries (i.e.

not entries 2, 5, 6) of *row* 1 of G, with reversed sign. The result is

$$H = \begin{bmatrix} \cdot & -2 & \cdot & \cdot & \cdot & \cdot & \cdot \\ \cdot & -0 & \cdot & \cdot & \cdot & \cdot & \cdot \\ \cdot & -4 & \cdot & \cdot & \cdot & \cdot & \cdot \\ \cdot & -1 & \cdot & \cdot & \cdot & \cdot & \cdot \end{bmatrix}.$$

Similarly, look at columns 5 and 6 of G. These have their 1s in positions 3 and 2 respectively, so columns 5 and 6 of H are occupied by the remaining entries of rows 3 and 2 of G with reversed signs. Finally, the columns of the 4×4 unit matrix are inserted as columns 1, 3, 4 and 8 of H. This gives

$$\begin{aligned} H &= \begin{bmatrix} 1 & -2 & 0 & 0 & -1 & -3 & 0 \\ 0 & -0 & 1 & 0 & -2 & -2 & 0 \\ 0 & -4 & 0 & 1 & -3 & -1 & 0 \\ 0 & -1 & 0 & 0 & -2 & -0 & 1 \end{bmatrix} \\ &= \begin{bmatrix} 1 & 3 & 0 & 0 & 4 & 2 & 0 \\ 0 & 0 & 1 & 0 & 3 & 3 & 0 \\ 0 & 1 & 0 & 1 & 2 & 4 & 0 \\ 0 & 4 & 0 & 0 & 3 & 0 & 1 \end{bmatrix} \end{aligned}$$

[Ex 35–38]

Now we have a method of getting from G to H quickly, but only if G has a rather special form. The fact which makes the method practically useful is that every linear code has a generator matrix of the special form, and that there is an easy way to find it. To establish this we need some ideas involved in continuing the equivalent code theme of Chapter 4 to linear codes in particular.

5.8 Equivalence of linear codes

Using the definition of equivalence established in Chapter 4 it is easy to see that one of a pair of equivalent codes may be linear and the other not. For example $C = \{0000, 1001, 0110, 1111\}$ is linear, and is equivalent to $C' = \{1000, 0001, 1110, 0111\}$ by a symbol change in the first position. But C' is not linear since $\mathbf{0} \notin C'$.

So we ask the question: is it possible to define a notion of equivalence *entirely within* the class of *linear* codes? Guided by Chapter 4, our aim is, given any linear code C, to define a set of 'equivalence operations', any sequence of which, applied to C, will produce an equivalent code C' *which is also linear.*

Of the two transformations used to define ordinary equivalence, we have seen in our initial example that symbol permutations are a problem. But the other, positional permutation, causes no difficulty as it is clear that a linear code remains linear if its positions are permuted. Furthermore, any permutation of the positions of C corresponds to the same permutation performed on the columns of its generator matrix G. This observation is important for the rest of this discussion.

Our aim now is to restrict the set of allowed symbol permutations so that linearity is preserved. Again our example provides guidance : for C' to be linear it is necessary that it contains $\mathbf{0}$. Now if $d(C) \geq 2$, so that the weight of any non-zero codeword is at least 2, then the result of doing any symbol permutation which does not fix the zero symbol, in any position, is a code which does not contain $\mathbf{0}$. The next two Exercises give a plausible solution to this problem, and reasons why it won't work! [Ex 39, 40]

So we are forced to restrict the allowed permutations and our choice of restriction is given by Exercise 20 of Chapter 3, which we can recast as follows: if p is prime and $0 < a < p$, then the sequence $0a, 1a, 2a, \cdots, (p-1)a$ is a permutation of $0, 1, 2, \cdots, p-1$. Clearly $0a = 0$ so it is also a permutation which fixes 0. It also preserves linearity. To see this, suppose C' is the result of multiplying the ith components of all codewords of C by a_i $(i = 1, 2, \cdots, n)$ and let $(\mathbf{c})_i, (\mathbf{c}')_i$ denote the ith components of codewords $\mathbf{c} \in C$ and the corresponding $\mathbf{c}' \in C'$. Then for any $\mathbf{c}_1', \mathbf{c}_2' \in C'$ we have

$$(\mathbf{c}_1' + \mathbf{c}_2')_i = (\mathbf{c}_1')_i + (\mathbf{c}_2')_i = a_i\,(\mathbf{c}_1)_i + a_i\,(\mathbf{c}_2)_i = a_i\,(\mathbf{c}_1 + \mathbf{c}_2)_i$$

and $\mathbf{c}_1 + \mathbf{c}_2 \in C$ so $\mathbf{c}_1' + \mathbf{c}_2' \in C'$. Closure under scalar multiplication is checked just as easily.

Unlike arbitrary symbol permutations, this restricted class, if done only to the columns of a generator matrix, is then 'inherited' by the whole code. To be precise about this, let C be a linear code with generator G. Let C'' be the code obtained by multiplying the ith components of all codewords of C by a $(\neq 0)$, and let C' be the code generated by G', the matrix obtained by multiplying the ith column of G by a. It is easy to check that $C' = C''$.

Now we can try to formulate our definition of equivalence within linear codes entirely in terms of operations on their generator matrices. We also need a name for the new concept of equivalence. For the moment we shall call it *linear equivalence*, but this name is not standard, and in common with most of the literature, when the context is clearly linear codes we shall drop the 'linear', and it will be understood that the restricted idea of equivalence is intended.

Let C be defined by a generator matrix G and let G' be the result of applying to G any sequence of: permutations of the columns and/or multiplication of any columns by non-zero constants. We have seen that the code C' generated by G' is linearly equivalent to C. [Ex 41]

But this is not the whole story. It is easy to find pairs of matrices G, G' which generate linearly equivalent codes but G' is not obtainable from G

by applying the operations above. A simple example is

$$G = \begin{bmatrix} 1 & 0 \\ 0 & 1 \end{bmatrix}, \quad G' = \begin{bmatrix} 1 & 0 \\ 1 & 1 \end{bmatrix}$$

These both generate Z_2^2, but clearly the operations mentioned so far can never convert G to G'. So the problem is to find a set S of operations on generator matrices, with the following property : given any linear code C and one of its generator matrices G, the set of matrices produced by applying all possible sequences of operations in S to G is precisely the set of all generator matrices of all codes linearly equivalent to (or the same as) C.

You may wonder why we should be interested in having matrices which generate *the same* code as the original, as well as codes merely equivalent to it. Part of the answer has already been given at the end of section 5.4 where we noted the convenience of standard form generators for decoding, and in section 5.7 where standard or 'nearly standard' form for G made the process of getting H much simpler.

Here are the operations on G which make up the set S.

R1 Permutation of the rows

R2 Replacement of a row by one of its non-zero multiples

R3 Replacement of a row by the sum of itself and any multiple of another row

C1 Permutation of the columns

C2 Replacement of a column by one of its non-zero multiples.

We have already discussed $C1$ and $C2$. These are the only two of the five which can change the code. The three row operations only change the form of G but leave the code unchanged. You are asked to prove this last claim now. [Ex 42]

I ask you to take on trust the fact that if matrices G, G' generate linearly equivalent codes then there is a sequence of operations of some or all of the five types which converts G to G'. This should come as no surprise to those readers who have used various reduction techniques on matrices in order to solve systems of linear equations. We make no formal use of the result, but only mention it here to shed further light on why these five operations are the vital ones.

At the end of section 5.7 we used the special form of matrix which we now define formally.

Definition 5.7 A $k \times n$ generator matrix is in *nearly standard* form if k of its columns are the k columns of I_k, the $k \times k$ unit (or identity) matrix.

Now we give an algorithm to demonstrate (rather than prove formally) the theorem alluded to at the end of section 5.7.

Theorem 5.14 Given a linear code C with generator matrix G, row operations $R1, R2$ and $R3$ suffice to construct a generator matrix G' for C, in nearly standard form.

The algorithm to achieve this is simply described as follows:

(a) select any non-zero member of row 1, say a_{ij}.

(b) multiply row 1 by a_{1j}^{-1} so that row 1 now has 1 in its jth place.

(c) for all $i \neq 1$ replace row i by row $i - a_{ij} \times$ row 1. The matrix now has column j equal to the first column of I_k. Repeat this for all rows.

Here is an example of the algorithm in action (over the field Z_5). The non-zero entry selected in each row is underlined and the type of operation used at each step is shown, followed by the details of the particular operations. After this we give an indication of why it always works.

$$G = \begin{matrix} 0 & \mathit{2} & 0 & 3 & 3 & 4 \\ 1 & 2 & 3 & 4 & 0 & 1 \\ 2 & 2 & 3 & 3 & 4 & 4 \end{matrix} \quad \underset{R2}{\overset{\text{STEP 1}}{\longrightarrow}} \quad \begin{matrix} 0 & 1 & 0 & 4 & 4 & 2 \\ 1 & 2 & 3 & 4 & 0 & 1 \\ 2 & 2 & 3 & 3 & 4 & 4 \end{matrix}$$

$$\underset{R3}{\overset{\text{STEP 2}}{\longrightarrow}} \quad \begin{matrix} 0 & 1 & 0 & 4 & 4 & 2 \\ \mathit{1} & 0 & 3 & 1 & 2 & 2 \\ 2 & 0 & 3 & 0 & 1 & 0 \end{matrix} \quad \underset{R3}{\overset{\text{STEP 3}}{\longrightarrow}} \quad \begin{matrix} 0 & 1 & 0 & 4 & 4 & 2 \\ 1 & 0 & 3 & 1 & 2 & 2 \\ 0 & 0 & 2 & 3 & 2 & \mathit{1} \end{matrix}$$

$$\underset{R3}{\overset{\text{STEP 4}}{\longrightarrow}} \quad \begin{matrix} 0 & 1 & 1 & 3 & 0 & 0 \\ 1 & 0 & 4 & 0 & 3 & 0 \\ 0 & 0 & 2 & 3 & 2 & 1 \end{matrix} = G'$$

step 1 row 1 replaced by 3 × row 1
step 2 row 2 → row 2 − 2 × row 1, row 3 → row 3 − 2 × row 1
step 3 row 3 → row 3 − 2 × row 2
step 4 row 1 → row 1 − 2 × row 3, row 2 → row 2 − 2 × row 3.
and columns 2, 1 and 6 are the required unit columns.

Since the initial G is a generator matrix, its rows are independent. Hence its first row must have a non-zero entry, so the algorithm can certainly produce the column with 1 in its first place and zero elsewhere. By Exercise 42 the matrix produced at each stage also generates C so its rows remain independent. Hence row 2 of the new matrix also has a non-zero entry, so the algorithm can be applied again, ⋯ and so on for all the rows. Finally note that each time part (c) of the algorithm is applied, this has no effect on columns already transformed to the required form because in these columns only multiples of zero are added to their entries! □

[Ex 43]

5.9 Erasure correction and syndromes

In section 2.3 we briefly mentioned the decoding of words with erasures. Now we look more closely at this problem, first directing our attention to channels which induce erasures but not errors.

Definition 5.8 A code of length n is called e-erasure decodable if for each word $\boldsymbol{r}$ received with f ($f \leq e$) erasures (but no other errors), there is a unique codeword which agrees with $\boldsymbol{r}$ at the other $n - f$ positions.

There is a simple connection between $d(C)$ and the erasure decodability of C, very similar to that between $d(C)$ and the error detecting capability of C given by Theorem 2.1.

Theorem 5.15 C is e-erasure decodable if and only if $d(C) \geq e + 1$.

Proof.

(i) Suppose C is not e-erasure decodable. Then there exists a codeword $\mathbf{c}$ and a set of $f (f \leq e)$ of its positions with the following property : $\mathbf{c}$ is transmitted, and is received with f of its symbols erased, and there is a codeword $\mathbf{c}'$ distinct from $\mathbf{c}$, which agrees with $\mathbf{c}$ at the other $n - f$ positions. Hence $d(\mathbf{c}, \mathbf{c}') \leq f \leq e$, so $d(C) \leq e$.

(ii) Conversely, suppose $d(C) \leq e$ and let $\mathbf{c}, \mathbf{c}'$ be distinct codewords with $d(\mathbf{c}, \mathbf{c}') = f \leq e$. Then the word $\boldsymbol{r}$ which has these f symbols erased but agrees with $\mathbf{c}$ at the other $n - f$ positions clearly agrees with $\mathbf{c}'$ at these positions, so C is not e-erasure decodable. □

The decoding strategy for an erasure channel and a code with $d(C) > e$ and a received word $\boldsymbol{r}$ with f erasures is simply to decode to a codeword $\mathbf{c}$ which agrees with $\boldsymbol{r}$ at the non-erased positions. If $f \leq e$ then by Theorem 5.15 $\mathbf{c}$ is unique. In cases where $f > e$ there may or may not be a unique $\mathbf{c}$ (see examples 2 and 3 below). For a large arbitrary code the scanning process could be very time-consuming, so let us see how using a linear code can help.

First some examples.

1. Let C be linear over Z_5 with

$$H = \begin{bmatrix} 3 & 3 & 4 & 1 & 0 \\ 0 & 2 & 4 & 0 & 1 \end{bmatrix}$$

and suppose the word $\boldsymbol{r} = x2304$(x unknown) is received via an erasure channel. $syn(\boldsymbol{r}) = (3x+3, 0)$, and recall that $\boldsymbol{r}$ is a codeword if and only if $syn(\boldsymbol{r}) = \mathbf{0}$, so we have to solve $(3x+3, 0) = (0, 0)$, or $3x + 3 \equiv 0 \bmod 5$. From Chapter 3 we know this has a unique solution, which is easy to spot in this case: $x = 4$. Hence we decode as 42304.

Note that $d(C) = 2$ because H has column $1 = 3 \times$ column 4, so by Theorem 5.15 we know C is 1-erasure decodable. It is not 2-erasure decodable so there will be some instances of two erasures which are not decodable uniquely.

2. Same C and H, $\boldsymbol{r} = x23y4$, $syn(\boldsymbol{r}) = (3x + y + 3, 0) = \mathbf{0}$ if and only if $(x, y) = (0, 2), (1, 4), (2, 1), (3, 3)$ or $(4, 0)$.
3. Same C and H, $\boldsymbol{r} = xy112$, $syn(\boldsymbol{r}) = (3x + 3y, 2y + 1) = \mathbf{0}$ if and only if $(x, y) = (3, 2)$.

So in example 3 $\boldsymbol{r}$ could be decoded uniquely but in example 2 the best we could do was to narrow the choice down to one of five possible transmitted codewords. [Ex 44]

If you suspected that the different outcomes of examples 2 and 3 are related to the fact that in example 2 the erasures were in a pair of positions corresponding to a dependent pair of columns of H, then your suspicion is well-founded.

Theorem 5.16 A word received with e erasures (and no errors) is uniquely decodable if and only if the corresponding columns of H are independent.

To prove this we need the following lemma from linear algebra.

Lemma 5.1 If $\boldsymbol{x}$ is an unknown vector, A is a known matrix and $\boldsymbol{b}$ is a known vector, then the set of all solutions of $\boldsymbol{x}A = \boldsymbol{b}$ is $\boldsymbol{x}_0 + S$ where $\boldsymbol{x}_0$ is any particular solution of this equation and S is the set of all solutions of the equation $\boldsymbol{x}A = \mathbf{0}$. (The notation $\boldsymbol{x}_0 + S$ is to be interpreted as for cosets.)

Proof. Let $\boldsymbol{y} \in \boldsymbol{x}_0 + S$ so $\boldsymbol{y} = \boldsymbol{x}_0 + \boldsymbol{s}$ for some $\boldsymbol{s} \in S$.

Then $\boldsymbol{y}A = \boldsymbol{x}_0A + \boldsymbol{s}A = \boldsymbol{b} + \mathbf{0} = \boldsymbol{b}$, so all members of $\boldsymbol{x}_0 + S$ are solutions of $\boldsymbol{x}A = \boldsymbol{b}$.

Conversely, let $\boldsymbol{z}$ be any solution of $\boldsymbol{x}A = \boldsymbol{b}$, so $\boldsymbol{z}A = \boldsymbol{b}$. Now $\boldsymbol{z} = \boldsymbol{x}_0 + (\boldsymbol{z} - \boldsymbol{x}_0)$ so $(\boldsymbol{x}_0 + (\boldsymbol{z} - \boldsymbol{x}_0))A = \boldsymbol{b}$. But the left hand side of this is

$$\boldsymbol{x}_0A + (\boldsymbol{z} - \boldsymbol{x}_0)A = \boldsymbol{b} + (\boldsymbol{z} - \boldsymbol{x}_0)A$$

so

$$\boldsymbol{z} - \boldsymbol{x}_0)A = \boldsymbol{o}$$

and

$$\boldsymbol{z} - \boldsymbol{x}_0 \ \in S.$$

Hence

$$\boldsymbol{z} \in \boldsymbol{x}_0 + S.$$

□

Proof of Theorem 5.16. Without loss of generality, and to make the notation easier to handle, let the erasures be in the first e positions.

Let $\mathbf{c} = c_1c_2 \cdots c_ec_{e+1} \cdots c_n$ be the transmitted codeword, which is received as ??⋯? $c_{e+1} \cdots c_n$. We take the received word to be $\boldsymbol{r} = 00 \cdots 0$ $c_{e+1} \cdots c_n$ with errors in some or all of the first e positions and none in the rest. This means that the error pattern $\boldsymbol{x} = x_1 \cdots x_e 0 \cdots 0$, and finding $\boldsymbol{x}$ is equivalent to finding $\mathbf{c}$ because

$$\boldsymbol{x} = \boldsymbol{r} - \mathbf{c} = -c_1 - c_2 \cdots - c_c 0 \cdots \ 0.$$

For each word $\boldsymbol{w}$ of length n we shall write $\boldsymbol{w}'$ for the word of length e consisting of the first e components of $\boldsymbol{w}$. By Theorem 5.10 $syn(\boldsymbol{x}) = syn(\boldsymbol{r})$.

That is, $x'H'^T = rH^T$ where H' is the matrix consisting of the first e columns of H. Since r and H are known, the right hand side is a known word which we shall now call b. So identifying the erasures is now equivalent to solving the system

$$x'H'^T = b$$

This certainly has a solution $x' = -c'$, so from the lemma, the set of all solutions is $-c' + S$ where S is the set of all solutions of $x'H'^T = \mathbf{0}$, and note that the left hand side of this is just a linear combination of the first e columns of H. If the columns of H' are independent this has only the trivial solution $\mathbf{0}'$, in which case c' is the only possible vector of erasures. If they are dependent, then there are more solutions, simply by the definition of dependence, and the received word is not decodable uniquely. □

[Ex 45, 46]

Now suppose the channel induces both errors and erasures. First we examine another example.

4. Same H as previous examples, $r = 123x2$. This time the syndrome is $(1 + x, 3)$ which can never be $\mathbf{0}$ so there must be at least one error. We make the usual reasonable working assumption of only one error and try to decode on this basis, aided by the fact that an error pattern $e0000$ will have a syndrome which is $e\times$ column 1 of H, and similarly for other error patterns of weight 1. We try all possibilities for x.

   $x = 0 \Rightarrow syn(r) = (13)$ which is not a multiple of a column of H
   $x = 1 \Rightarrow syn(r) = (23)$ which is $4\times$ column 2
   $x = 2 \Rightarrow syn(r) = (33)$ which is $2\times$ column 3
   $x = 3 \Rightarrow syn(r) = (43)$ which is not a multiple of a column of H
   $x = 4 \Rightarrow syn(r) = (03)$ which is $3\times$ column 5

   Hence there are just three possibilities consistent with a single error

$$\begin{array}{lcccccccc} & c & = & 12312 & - & 04000 & = & 13312 \\ \text{or} & c & = & 12322 & - & 00200 & = & 12122 \\ \text{or} & c & = & 12342 & - & 00003 & = & 12344 \end{array}$$

[Ex 47]

Notice that when we regard the codes of example 4 and Exercise 47 purely as error correcting codes the former is 0-error correcting (d is only 2) but the latter has $d = 3$ so is 1-error correcting . It is curious then, that for these particular instances of a single error combined with a single erasure, the better pure error correcting code behaves worse than the poorer one in terms of their ability to narrow down the range of possibilities. [Ex 48, 49]

Finally we return to the case of an arbitrary block code and find a necessary and sufficient condition for received words r with erasures and errors to be correctly decoded. The job of the decoder is to fill in the erased positions of $\boldsymbol{r}$ so that the resulting word $\boldsymbol{r}'$ agrees with some codeword $\mathbf{c}'$ at the erased positions and from all the candidate pairs $(\boldsymbol{r}', \mathbf{c}')$ pick one say $(\boldsymbol{r}^*, \mathbf{c}^*)$ which minimizes $d(\boldsymbol{r}', \mathbf{c}')$, and decode $\boldsymbol{r}$ to $\mathbf{c}^*$.

We define a code to be t/e-error/erasure decoding if for any received word with at most t errors and at most e erasures, the result of the process described above is the transmitted codeword and no other codeword.

Theorem 5.17 C is t/e-error/erasure decoding if and only if $d(C) \geq 2t + e + 1$.

Proof. Let C be a code which is not t/e-error/erasure decoding. Then there is a codeword $c = a_1 a_2 \ldots a_m a_{m+1} \ldots a_n$ which is received as r, where $r = ?? \ldots ?a''_{m+1} \cdots a''_n$, and r is not uniquely decoded to c. r has $m (\leq e)$ erasures and at most t errors in the non-erased positions. Assuming that the m erasures are in the first m positions is a convenience which does not affect the generality of the argument.

Then the fact that r is not uniquely decoded to c means that there must be a pair of choices of words r_1 and r_2 agreeing with r at the non-erased places, and another codeword $c' = a'_1 \ldots a'_n$, also differing from r in at most t of the non-erased places, such that

$$d(c', r_1) \leq d(c, r_2)$$

$$\begin{aligned}
\text{Then } d(c, c') &= \text{number of disagreements in the first } m \text{ places} \\
&\quad + \text{number of disogreements in the rest} \\
&\leq m + d(a_{m+1} \ldots a_n, a'_{m+1} \cdots a'_n) \\
&\leq e + d(a_{m+1} \ldots a_n, a''_{m+1} \cdots a''_n) \\
&\quad + d(a''_{m+1} \cdots a''_n, a'_{m+1} \cdots a'_n) \\
&\quad \text{(by the triangle inequality)} \\
&\leq e + t + t. \\
\text{Hence } d(C) &\leq e + 2t.
\end{aligned}$$

Conversely, suppose $d(C) < 2t+e+1$. We show C is not t/e-error/erasure decoding.

Let $d(C) = d < 2t + e + 1$, and let $\mathbf{c}, \mathbf{c}'$ be codewords with $d(\mathbf{c}, \mathbf{c}') = d$. For simplicity we take $\mathbf{c}$ and $\mathbf{c}'$ to differ in their first d places. Now suppose $\mathbf{c}$ is sent and is received as $\boldsymbol{r}$, a word having $e' (\leq e)$ erasures and $t' (\leq t)$ errors in the non-erased positions. There are three cases:

(1) $t' > d$;

(2) $t' \leq d < t' + e'$;

(3) $t' + e' \leq d$.

For these cases respectively, take the forms for $\boldsymbol{r}$ shown below with $\mathbf{c}$ and $\mathbf{c}'$.

$$\begin{array}{lllllll}
\mathbf{c} & = a_1 \cdot\cdot & \cdot\cdot\cdot\cdot & \cdot\, a_d\ a_{d+1} \cdot\cdot & \cdot\cdot\cdot & \cdot\, a_n & \\
\mathbf{c}' & = b_1 \cdot\cdot & \cdot\cdot\cdot\cdot & \cdot\, b_d\ a_{d+1} \cdot\cdot & \cdot\cdot\cdot & \cdot\, a_n & \\
\boldsymbol{r} & = b_1 \cdot\cdot & \cdot\cdot\cdot\cdot & \cdot\, b_d\ x_{d+1} \cdot x_{t'}, & ?\ ?\ a_{t'+e'+1} & \cdot\, a_n & \text{for case 1} \\
\boldsymbol{r} & = b_1 \cdot b_{t'} & ? \cdot\cdot\cdot & ?\ ?\ \ ? \quad \cdot\cdot & \cdot\ ?\ a_{t'+e'+1} & \cdot\, a_n & \text{for case 2} \\
\boldsymbol{r} & = b_1 \cdot b_{t'} & ? \cdot ?\ a_{t'+e'+1} & \cdot\, a_d\ a_{d+1} \cdot\cdot & \cdot\cdot\cdot & \cdot\, a_n & \text{for case 3}
\end{array}$$

and in case 3 also choose t' and e' such that

$$t' = \begin{cases} \dfrac{d}{2} & \text{if } d \text{ is even} \\[2ex] \dfrac{d+1}{2} & \text{if } d \text{ is odd.} \end{cases}$$

Let the replacements for the erasures be chosen to coincide with the $\mathbf{c}'$ symbols, then in case (1) $d(\boldsymbol{r}, \mathbf{c}') = t' - d$ but $d(\boldsymbol{r}, \mathbf{c}) = t'$; in case (2) $d(\boldsymbol{r}, \mathbf{c}') = 0$; in case (3) $d(\boldsymbol{r}, \mathbf{c}') = d - t' - e'$ and $d(\boldsymbol{r}, \mathbf{c}) = t' + e'$. In this last case $d(\boldsymbol{r}, \mathbf{c}) - d(\boldsymbol{r}, \mathbf{c}') = 2t' + 2e' - d$. For d even this is $d + 2e' - d \geq 0$, and for d odd it is $d + 1 + 2e' - d > 0$. So in all cases there is an $\boldsymbol{r}$ which nearest neighbour decoding would not uniquely decode to $\mathbf{c}$.

5.10 Exercises for Chapter 5

1. For the field Z_2 and $\boldsymbol{u} = 1011001, \boldsymbol{v} = 1101010, \alpha = 0, \beta = 1$ work out $\boldsymbol{u} + \boldsymbol{v}, \boldsymbol{u} - \boldsymbol{v}, -\boldsymbol{u}, \alpha\boldsymbol{v}, \beta\boldsymbol{v}$. For the field Z_5 and $\boldsymbol{u} = 2033004, \boldsymbol{v} = 1402041, \alpha = 3, \beta = 4$ work out $\boldsymbol{u} - \boldsymbol{v}$ and $\alpha\boldsymbol{u} + \beta\boldsymbol{v}$.
2. Show that conditions (i), (ii) in Definition 5.2 are equivalent to the single condition : for all $\mathbf{c}, \mathbf{c}'$ in C and all α, β in $Z_p, \alpha\mathbf{c} + \beta\mathbf{c}' \in C$.
3. Show that for a *binary* linear code condition (ii) in Definition 5.2. may be omitted.
4. For any (not necessarily linear) code over Z_p, show that $d(\mathbf{c}_1, \mathbf{c}_2) = w(\mathbf{c}_1 - \mathbf{c}_2)$.
5. Which of the following binary codes are linear? Find their minimum distances. $\{101, 111, 011\}$, $\{000, 001, 010, 011\}$, $\{0000, 0001, 1110\}$, $\{00000, 11100, 00111, 11011\}$, $\{00000, 11110, 01111, 10001\}$, $\{000000, 101010, 010001, 111111\}$.
6. Show that in a linear binary code either the first bit of every codeword is 0 or exactly half the codewords begin with 0.
7. Let S be a non-empty set of words in Z_p^n. Show that $\langle S \rangle$ is linear.
8. Find a non-zero word in Z_5^6 orthogonal to 123142.
9. Show that every word of even weight in a binary code is orthogonal to itself, and any two words of the same weight have even distance.

10. Find $S^\perp$ and $T^\perp$ for $S = \{1202, 1111, 2000\} \subseteq Z_3^4$ and $T = \{10001, 00111, 11000, 01110\} \subseteq Z_2^5$.
11. Show that $S^\perp$ is a linear code irrespective of whether S is linear or not.
12. Let $\{\boldsymbol{c}_1, \cdots, \boldsymbol{c}_m\}$ be an independent set of words in Z_p^n and let $\alpha_i, \beta_i (i = 1, 2, \cdots, m)$ be members of Z_p. If $\sum_{i=1}^m \alpha_i \boldsymbol{c}_i = \sum_{i=1}^m \beta_i \boldsymbol{c}_i$ show that, for all i, $\alpha_i = \beta_i$.
13. Find the spans of the following sets of binary words

    (i) $\{1010, 0101, 1111\}$,
    (ii) $\{0101, 1010, 1100\}$,
    (iii) $\{10101, 00111, 01011, 11001\}$, and
    (iv) the ternary words $\{1011, 0112\}$.

14. Find bases for the spans of the following sets:

    (i) $\{1100, 1010, 0000, 1001, 0101\}$ over Z_2 and
    (ii) $\{0140, 4322, 1000, 1234, 3410\}$ over Z_5.

    Extend the second basis to a basis of Z_5^4.
15. Convince yourself that the claims made in (10) are correct.
16. If a linear code over Z_p has dimension k, how many codewords does it have? [Hint : Exercise 12 will help]
17. Prove result (11) of section 5.3. [Hint : use (9)]
18. Check that the method of encoding described here ensures that no pair of distinct messages are encoded to the same codeword.
19. Spot a codeword of weight 3 in the example of section 5.4.
20. Check that if you choose any words $\boldsymbol{x}, \boldsymbol{y}$ from the second and third rows respectively, then $C + 222 = C + \boldsymbol{x}$ and $C + 200 = C + \boldsymbol{y}$ (the only thing which changes is the order in which the words of a row appear).
21. Prove that for any $[n, k]$ code C over Z_p:

    (i) all cosets have the same size;
    (ii) $C + \boldsymbol{x} = C + \boldsymbol{y}$ if $y \in C + \boldsymbol{x}$, and $(C + \boldsymbol{x}) \cap (C + \boldsymbol{y}) = \phi$ if $\boldsymbol{y} \notin C + \boldsymbol{x}$;
    (iii) Every word of Z_p^n is a member of some coset;
    (iv) $\boldsymbol{x}, \boldsymbol{y}$ are in the same coset if and only if their difference is in C;
    (v) there are p^{n-k} distinct cosets.

22. The binary linear code C and the ternary linear code C' have generator matrices

$$\begin{bmatrix} 1 & 0 & 0 & 1 & 1 \\ 0 & 1 & 0 & 1 & 1 \\ 0 & 0 & 1 & 0 & 1 \end{bmatrix} \quad \text{and} \quad \begin{bmatrix} 0 & 1 & 2 & 1 \\ 1 & 2 & 2 & 0 \end{bmatrix}$$

respectively. Construct a Slepian array for C and use it to decode 01100. For C' list the codewords and state the number of cosets.

23. For C of Exercise 22 find a pair of words $(\boldsymbol{c}, \boldsymbol{r})$ such that $d(\boldsymbol{c}, \boldsymbol{r}) = 1$ and if $\boldsymbol{c}$ is sent and $\boldsymbol{r}$ received, $\boldsymbol{r}$ is not correctly decoded. For C' explain why every word of weight 1 must be a coset leader, and why there are no coset leaders with weight greater than 1.

24. Let C be the set of all even weight words in Z_2^n. Show that C is a linear code. What is $C^{\perp}$? Find standard form generator matrices for C and for $C^{\perp}$.

25. Show that in a binary linear code C all words have even weight or half of them have even weight. If C has a generator matrix in which all the rows are of even weight show that the first of these holds.

26. C_1, C_2 are $[n_1, k, d_1], [n_2, k, d_2]$ codes generated by G_1, G_2 respectively. G is the matrix $[G_1|G_2]$ formed by writing G_2 to the right of G_1. If C is the code generated by G what can you deduce about $d(C)$?

27. Let $C + \boldsymbol{a}$ be any coset of a binary linear code C. Show that $C \cup (C + \boldsymbol{a})$ is a linear code.

28. Describe the Hamming 8-bit code of Chapter 2 as $S^{\perp}$ for a suitable set of words S.

29. If C is a linear code prove that $(C^{\perp})^{\perp} = C$.

30. Show that any repetition code over Z_p is linear and answer the question of Exercise 28 for such a code.

31. Find a parity check matrix for the linear code C over Z_3 with a generator matrix,
$$G = \begin{bmatrix} 1 & 1 & 1 & 0 \\ 2 & 0 & 1 & 1 \end{bmatrix}.$$

32. Find $d(C)$ if C has the generator matrix
$$\left[\begin{array}{c|cccc} & 1 & 1 & 0 & 0 \\ & 1 & 0 & 1 & 0 \\ & 0 & 1 & 1 & 0 \\ I_7 & 1 & 1 & 1 & 1 \\ & 1 & 1 & 0 & 1 \\ & 0 & 1 & 0 & 1 \\ & 1 & 0 & 0 & 1 \end{array}\right]$$
where I_7 is the 7×7 identity matrix.

33. Show that if C is a 1-error-correcting linear binary code, then no column of its p.c. matrix is $\mathbf{0}$ and no two columns are the same.

34. In the example under discussion show that no word of weight ≤ 2 has syndrome 1111. What does this tell you about the correctable errors?

35. Attempt to construct a parity check matrix for a $[6, 3, 4]$ binary code, and hence show that no such code can exist.

36. Find the minimum distances of the codes given by these parity check matrices.

(a)

$$\begin{bmatrix} 1 & 0 & 0 & 1 & 0 & 0 & 0 & 1 & 1 & 0 & 1 & 1 & 0 & 1 \\ 0 & 0 & 1 & 0 & 1 & 0 & 1 & 1 & 1 & 0 & 0 & 1 & 1 & 1 \\ 0 & 0 & 1 & 0 & 0 & 1 & 0 & 0 & 0 & 1 & 1 & 1 & 0 & 1 \\ 0 & 1 & 0 & 1 & 0 & 0 & 1 & 1 & 0 & 1 & 0 & 1 & 1 & 0 \end{bmatrix} \quad \text{over } Z_2,$$

(b)

$$\begin{bmatrix} 1 & 0 & 0 & 0 & 1 & 0 & 1 & 1 \\ 0 & 1 & 0 & 0 & 0 & 1 & 1 & 1 \\ 0 & 0 & 1 & 0 & 1 & 1 & 0 & 1 \\ 0 & 0 & 0 & 1 & 1 & 1 & 1 & 0 \end{bmatrix} \text{over} Z_2,$$

(c)

$$\begin{bmatrix} 2 & 1 & 0 & 6 & 1 & 0 & 0 \\ 0 & 1 & 3 & 0 & 4 & 3 & 2 \\ 4 & 0 & 4 & 6 & 0 & 3 & 5 \end{bmatrix} \quad \text{over } Z_7,$$

(d)

$$\begin{bmatrix} 1 & 1 & 0 & 0 & 0 & 1 & 0 \\ 0 & 1 & 1 & 1 & 0 & 0 & 0 \\ 0 & 1 & 0 & 1 & 0 & 1 & 1 \end{bmatrix} \quad \text{over} Z_{31}.$$

37.

$$G = \left[\begin{array}{c|ccc} & 1 & 1 & 1 \\ & 1 & 1 & 0 \\ I_4 & 1 & 0 & 1 \\ & 0 & 1 & 1 \end{array} \right].$$

Given that G is a generator matrix for a perfect [7, 4, 3] binary code, construct a syndrome table and use it to decode:

$$0000011, 1111111, 1100110$$

38.

$$\begin{bmatrix} 1 & 1 & 1 & 0 \\ 2 & 0 & 1 & 1 \end{bmatrix}$$

is a generator matrix for a ternary code C. Find a parity check matrix for C and use syndrome decoding to decode 2121, 1201 and 2222.

39. Suppose that instead of restricting the class of permutations we restrict instead what they act upon. Specifically, take an arbitrary symbol permutation π. From a code C generated by G define a new matrix G' which is just G with the entries of the jth column, $g_{ij} (i = 1, 2, \cdots, k)$ replaced by $\pi(g_{ij})$, and define C' as the linear code generated by G'.

For

$$G = \begin{bmatrix} 2 & 0 & 1 \\ 1 & 1 & 2 \end{bmatrix}$$

over Z_3 do

$$\pi = \begin{pmatrix} 0 & 1 & 2 \\ 2 & 1 & 0 \end{pmatrix}$$

at position 3 and show that C' is *not* equivalent to C. Note also that C' is *not* the result of applying π to the 3rd position of C.

40. With the same notation as in Exercise 39, do π on position 2 of G and show that things go even more drastically wrong.

41. Show by examples that in general both types of column operation produce codes which differ from C.

42. Show that if G generates C and G' is the result of applying any operation of type R1, R2 or R3 to G, then G' also generates C.

43. Apply the algorithm to convert the following to nearly standard form

$$G_1 = \begin{bmatrix} 1 & 1 & 1 & 1 & 1 & 0 & 0 & 1 & 1 \\ 1 & 0 & 1 & 0 & 1 & 0 & 1 & 0 & 1 \\ 1 & 1 & 0 & 0 & 0 & 1 & 1 & 1 & 1 \end{bmatrix} \text{over} Z_2.$$

$$G_2 = \begin{bmatrix} 2 & 2 & 0 & 0 & 0 & 0 & 0 & 0 \\ 1 & 1 & 1 & 2 & 1 & 1 & 1 & 2 \\ 0 & 1 & 2 & 0 & 1 & 2 & 0 & 1 \end{bmatrix} \text{over} Z_3.$$

44. A ternary code has

$$H = \begin{bmatrix} 2 & 1 & 2 & 1 & 1 & 0 \\ 1 & 1 & 2 & 1 & 0 & 1 \\ 0 & 1 & 0 & 2 & 0 & 0 \end{bmatrix}.$$

Decode the received words $1xyz12$ and $xyz210$.

45. Why can a linear code of length 10 and dimension 6 never uniquely decode words with 5 erasures?

46. Two linear codes over Z_5 have parity check matrices

$$\begin{bmatrix} 1 & 3 & 1 & 2 & 4 \\ 2 & 4 & 1 & 1 & 2 \end{bmatrix} \text{and} \begin{bmatrix} 3 & 2 & 1 & 0 & 4 \\ 4 & 2 & 1 & 3 & 2 \end{bmatrix}$$

Which code is better with respect to 2-erasure decodability?

47. Do a similar analysis for

$$H = \begin{bmatrix} 3 & 3 & 4 & 1 & 0 \\ 1 & 2 & 4 & 0 & 1 \end{bmatrix}$$

and $\boldsymbol{r} = x4423$.

48. Show that the binary code with

$$H = \begin{bmatrix} 1 & 0 & 1 & 1 & 0 & 0 \\ 1 & 1 & 1 & 0 & 1 & 0 \\ 0 & 1 & 1 & 0 & 0 & 1 \end{bmatrix}$$

will determine the transmitted word uniquely, on the assumption of at most one error, if $\boldsymbol{r} = 10110x$. Demonstrate this result directly by using the list of codewords.

49. Decode the received ternary words $1x20y1, 21xy11$ for a code with

$$H = \begin{bmatrix} 2 & 1 & 2 & 1 & 1 & 0 \\ 1 & 1 & 2 & 1 & 0 & 1 \\ 0 & 1 & 0 & 2 & 0 & 0 \end{bmatrix}.$$

6

The Hamming family and friends

6.1 Introduction

This chapter, like the previous one, mainly explores further consequences of linearity. The Hamming codes we met in Chapter 2 are but two members of a family of codes, all with pleasant useful properties, and as we shall see, other important codes can be constructed from the Hamming family. These other codes also appear at the ends of totally different lines of argument, but that is one of the delights of coding theory – the variety of interesting routes from A to B.

6.2 Hamming codes

These are most conveniently defined by their parity check matrices, and designed, using Theorem 5.12, to be one-error correcting. The Hamming codes form a 2-parameter family which we now define.

Definition 6.1 Ham(r,q) is the set of all linear [n,k] codes over Z_q whose p.c. matrices H have r rows and n columns, where n is the greatest possible number of columns consistent with the condition that no pair of columns are dependent.

This definition implies $k = n - r$, but what is n? The next theorem answers this and also provides a method of constructing a suitable H.

Theorem 6.1 All codes in Ham(r,q) have length

$$n = \frac{q^r - 1}{q - 1}$$

Proof. The condition on the columns in the definition above is equivalent to saying that no column is a multiple of any other. For each non-zero r-tuple $\boldsymbol{u}$, let $m(\boldsymbol{u})$ be the set of all its non-zero multiples, so $m(\boldsymbol{u})$ has $q-1$ members. There are $q^r - 1$ non-zero r-tuples in all. Now suppose $m(\boldsymbol{u})$ and

$m(\boldsymbol{v})$ have a member in common, say $\boldsymbol{x}$, and let $\boldsymbol{a}$ be *any* member of $m(\boldsymbol{u})$. Then $\boldsymbol{a} = \alpha\boldsymbol{u}$ and $\boldsymbol{x} = \beta\boldsymbol{u} = \gamma\boldsymbol{v}$ for some non-zero α, β, γ, and

$$\boldsymbol{a} = \alpha\beta^{-1}\boldsymbol{x} = \alpha\beta^{-1}\gamma\boldsymbol{v} \in m(\boldsymbol{v})$$

So $m(\boldsymbol{u}) \subseteq m(\boldsymbol{v})$, and by an identical argument $m(\boldsymbol{v}) \subseteq m(\boldsymbol{u})$, so $m(\boldsymbol{u}) = m(\boldsymbol{v})$. This means that if $m(\boldsymbol{u})$ and $m(\boldsymbol{v})$ are not identical they are totally disjoint, so that the distinct $m(\boldsymbol{u})$s partition the set of $q^r - 1$ non-zero r-tuples into subsets, each of size $q - 1$. Hence there are $\frac{q^r-1}{q-1}$ subsets, with the property that if we select one member from each subset, and make the selected members the columns of H, then H will have no pair of dependent columns. This is the *best* we can do because any H with more than this number of columns must have a pair of columns from the same subset, and these will be dependent. □

[Ex 1]

For large codes, doing the partition described above is not feasible, but there are ways of selecting the columns of H without doing this. One way is to select all those non-zero r-tuples whose first non-zero symbol is 1. Exercise 2 asks you to show that this works. [Ex 2]

Theorem 6.2 For given r,q, all codes of Ham(r,q) are linearly equivalent.

Proof. Let H be any p.c. matrix of any code in Ham(r,q). We know from the proof of the previous theorem that the columns of H must be a selection of one from each of the sets $m(\boldsymbol{u})$, and any pair from a fixed $m(\boldsymbol{u})$ are multiples of each other. So whichever selection is made it can be converted to any other by multiplying each column by the appropriate constant. □

[Ex 3]

By the way Ham(r,q) was constructed we know that all its codes have a minimum distance of at least of 3. In fact it is exactly 3.

Theorem 6.3 All Hamming codes have a minimum distance of 3.

Proof. Let H be a p.c. matrix for $C \in$ Ham(r,q). The three columns of H whose existence was established in Exercise 3 are dependent since

$$a^{-1}c\boldsymbol{x} + b^{-1}c\boldsymbol{y} + (-1)\boldsymbol{z} = \boldsymbol{o}$$

Then Theorem 5.12 gives the stated result and establishes that all Hamming codes are 1-error-correcting. □

Theorem 6.4 All Hamming codes are perfect.

Proof. We know now that for $C \in$ Ham(r,q), C is an [n,k,d] code where $d = 3$,

$$n = \frac{q^r - 1}{q - 1} \quad \text{and} \quad k = \frac{q^r - 1}{q - 1} - r.$$

Checking the Hamming bound:

$$\frac{q^n}{\sum_{i=0}^{1}\binom{n}{i}(q-1)^i} = \frac{q^n}{1+n(q-1)} = \frac{q^n}{1+(q^r-1)} = q^{n-r} = q^k = M$$

□

6.3 Decoding Ham(r,q)

Syndrome decoding is easy, and it can be made easier by taking the columns of H to be those suggested immediately after the proof of Theorem 6.1 and then, regarding each column $(a_1a_2\ldots a_r)^{\mathrm{T}}$ as representing the base q number $a_1q^{r-1}+a_2q^{r-2}+\ldots+a_r$, ordering the columns in increasing order of these numbers. We illustrate for Ham(3,5).

The H specified above is

$$\begin{bmatrix} 0\,0\,0\,0\,0\,0\,1\,1\,1\,1\,1\,1\,1\,1\,1\,1\,1\,1\,1\,1\,1\,1\,1\,1\,1\,1\,1\,1\,1\,1\,1 \\ 0\,1\,1\,1\,1\,1\,0\,0\,0\,0\,0\,1\,1\,1\,1\,1\,2\,2\,2\,2\,2\,3\,3\,3\,3\,3\,4\,4\,4\,4\,4 \\ 1\,0\,1\,2\,3\,4\,0\,1\,2\,3\,4\,0\,1\,2\,3\,4\,0\,1\,2\,3\,4\,0\,1\,2\,3\,4\,0\,1\,2\,3\,4 \end{bmatrix}$$

Now each non-zero syndrome will be a multiple of one column of H. For example, if the received word $\boldsymbol{r}$ has syndrome 341, this is 3×132 and 132 is the 24th column of H, so we simply subtract 3 from the 24th symbol of $\boldsymbol{r}$ to obtain the decoded word. The numerical ordering of the columns makes it easier to locate the column $(132)^{\mathrm{T}}$ in H, just as alphabetical ordering makes it easier to find a word in a dictionary!

For *binary* Hamming codes the process is still easier because any distinct pair of non-zero columns are independent so a p.c. matrix for Ham(r,2) must consist of all the non-zero binary strings of length r, and if ordered as described above the decoding process is: calculate syn($\boldsymbol{r}$); calculate the number i represented by syn($\boldsymbol{r}$) in binary; if $i=0$ assume there is no error, and if not change the ith bit of $\boldsymbol{r}$ to obtain the decoded word. [Ex 4–6]

6.4 Simplex codes

Temporarily forgetting about codes, consider ordinary Euclidean space and the problem of finding sets of points with the property that every pair of points in the set are separated by the same distance. Solutions are not very numerous: in two dimensions the only possibilities are a single pair of points or three points sitting at the vertices of an equilateral triangle; going to three dimensions only gives the extra solution of the four vertices of a regular tetrahedron. Hamming space is more interesting: one way of generating an equidistant set of words is strongly related to Hamming codes. First look again at the suggested solution to Exercise 6. You should find that whichever row operation you used you ended up with an H matrix all of whose rows have weight 4. Furthermore, all linear combinations of rows of H, except o, have weight 4. From this it follows that the dual of this Hamming code is an equidistant code. These facts are proved and set in a more general context in the following two theorems.

Theorem 6.5 In any linear code the distribution of codeword weights is identical to the distance distribution.

Proof. Let linear code C have M codewords and let A_w be the number of codewords of weight w. There are M^2 ordered pairs of codewords. Let $\mathbf{c}$ be any codeword of weight w and let $C = \{\mathbf{c}_1, \mathbf{c}_2, \ldots, \mathbf{c}_M\}$. Then $(\mathbf{c}_1, \mathbf{c}_1 - \mathbf{c}), (\mathbf{c}_2, \mathbf{c}_2 - \mathbf{c}), \ldots, (\mathbf{c}_M, \mathbf{c}_M - \mathbf{c})$ are M distinct ordered pairs of codewords each with $d(\mathbf{c}_i, \mathbf{c}_i - \mathbf{c}) = w$. Hence C has A_w codewords of weight w if and only if C has MA_w ordered pairs of codewords separated by distance w. □

Definition 6.2 The dual of any Hamming code is called a simplex code.

Theorem 6.6 All simplex codes are equidistant codes.

Proof. Let $C \in \text{Ham}(r,q)$ and let H be any one of its p.c. matrices. Then H generates $C^\perp$ and $C^\perp$ has $M = q^r$ codewords. Of the M^2 ordered pairs of codewords, M of them clearly have distance zero, and we have to show that the remaining $M^2 - M$ pairs have the same non-zero separation. Because of Theorem 6.5 this is equivalent to showing that the $M - 1$ non-zero words of $C^\perp$ have the same weight, which we prove is q^{r-1}. By Theorem 6.2 it suffices to consider any one of the equivalent codes in $\text{Ham}(r,q)$.

Now suppose $C^\perp$ has a codeword of weight $> q^{r-1}$, and choose a generator matrix G for $C^\perp$ in which this word is the first row. For those columns of G which start with a non-zero symbol, multiply these columns by the inverse of their first members so that the resulting matrix G′ generates a code equivalent to $C^\perp$, in which more than q^{r-1} columns start with 1. There are only q^{r-1} distinct ways of filling in the remaining entries of these columns, so they must include a repeated pair. Such a pair is of course dependent, which contradicts, via Theorems 5.9 and 5.12 the fact that C is one-error correcting.

If $C^\perp$ were to have a non-zero codeword of weight $< q^{r-1}$ a similar contradiction would be obtained by taking a generator for $C^\perp$ in which this word was the first row and then considering these columns (more than q^{r-1} of them) which start with zero. □

[Ex 7, 8]

We have seen examples of codes which are perfect and codes which are maximum distance separable (those which respectively meet the Hamming and Singleton bounds precisely). What about codes which meet the Plotkin bound? These codes exist but do not have a special name. But they do have a nice symmetry property:

Theorem 6.7 A binary code C satisfies the Plotkin bound if and only if it satisfies:

(i) C is equidistant, and

(ii) in each position exactly half the codewords have a 0.

Proof. Let C satisfy (i) and (ii). Then by (i) the inequality $S \geq (M^2 - M)d$ in the proof of Theorem 4.14 becomes equality:

$$S = (M^2 - M)d.$$

In the same proof, z_k becomes the constant $\frac{M}{2}$ by virtue of (ii) so

$$S = \frac{nM^2}{2}.$$

From these two equations it follows that

$$M = \frac{2d}{2d - n}.$$

Conversely, if C is a code satisfying

$$M = \frac{2d}{2d - n}$$

it follows that

$$(M^2 - M)d = \frac{nM^2}{2},$$

so the inequalities

$$(M^2 - M)d \leq S \leq \frac{nM^2}{2}$$

from the proof of Theorem 4.14 become equalities, and from the argument used to obtain the inequalities, this can only be the case if (i) and (ii) hold. □

[Ex 9]

6.5 Optimal linear codes

Note that by Theorem 5.4 or Exercise 6 of Chapter 5, the second condition of the previous theorem is almost redundant if we restrict ourselves to linear codes. We then have the following result.

Theorem 6.8 A linear binary code satisfies the Plotkin bound if and only if it is equidistant and there is no all-zero column in its generator matrix. □

The only lower bound we consider is Gilbert–Varshamov, but there is a problem if we try to apply the version proved in Chapter 4 to linear codes. Recall that this bound tells us that for given q, n, d there is a code with size at least that given by the G–V bound. But is there a *linear* code of at least that size? The next result answers that question.

Theorem 6.9 The linear Gilbert–Varshamov bound.

For any n, d and prime q there is a linear code with these parameters and size

$$M \geq \frac{q^n}{\sum_{i=0}^{d-1}\binom{n}{i}(q-1)^i}$$

Proof. We sequentially select words of Z_q^n to be codewords of the required code as follows. First, select $\mathbf{c}$, with $w(\mathbf{c}_1) \geq d$. If spheres of radius $d-1$ centred on the words in $< \mathbf{c}_1 >$ do not cover Z_q^n select $\mathbf{c}_2$ as one of the words not covered. If spheres of radius $d-1$ centred on the words in $< \mathbf{c}_1, \mathbf{c}_2 >$ do not cover Z_q^n select $\mathbf{c}_3$ as one of the words not covered $\cdots$ and so on $\cdots$ until Z_q^n is covered. If $\mathbf{c}_k$ is the last word to be selected we claim that the linear code $< \mathbf{c}_1, \mathbf{c}_2, \cdots, \mathbf{c}_k >$ has minimum distance $\geq d$ and a size which satisfies the inequality given by the theorem.

The size claim is proved by exactly the same argument as in the previous version of the G–V bound. To show that the distance is at least d we use induction.

Our induction hypothesis is that for some $i \geq 1, d(< \mathbf{c}_1, \cdots, \mathbf{c}_i >) \geq d$. This is clearly true of $i = 1$ because $\mathbf{c}$ has at least d non-zero symbols, so the same is true of all the non-zero multiples of $\mathbf{c}_1$.

For the inductive step, let C, C' be $< \mathbf{c}_1, \cdots, \mathbf{c}_i >$, $< \mathbf{c}_1, \cdots, \mathbf{c}_{i+1} >$ respectively. Let $\mathbf{c}'$ be any word of C' so that $\mathbf{c}' = \mathbf{c} + \alpha\mathbf{c}_{i+1}$ for some $\mathbf{c} \in C$ and some $\alpha \in Z_p$. If $\alpha = 0$ then $w(\mathbf{c}') = w(\mathbf{c}) \geq d$ by the induction hypothesis and the fact that C is linear. If $\alpha \neq 0$ then we have $w(\mathbf{c}') = w(\mathbf{c} + \alpha\mathbf{c}_{i+1}) = w(-\alpha^{-1}\mathbf{c} - \mathbf{c}_{i+1})$ (since multiplying a word by the non-zero constant $-\alpha^{-1}$ does not change its weight) $= d(-\alpha^{-1}\mathbf{c}, \mathbf{c}_{i+1}) \geq d$ by construction. This completes the induction so $< \mathbf{c}_1, \cdots, \mathbf{c}_k >$ has distance $\geq d$ as claimed. □

[Ex 10, 11]

One of the most important properties of Ham(r, q) is the perfection of all its codes. So what other perfect linear codes exist? Our first answer reinforces the view that Hamming codes are rather special.

Theorem 6.10 The only non-trivial linear perfect one-error correcting codes are the Hamming codes.

Proof. Let C be perfect, linear, one-error correcting, with alphabet Z_q for some prime q, having M codewords. Then

$$\begin{aligned}
M &= q^n / \sum_{i=0}^{1} \binom{n}{i} (q-1)^i = q^n/1 - n + qn \qquad (1)\\
&\Rightarrow\ q^n = M(1 - n + qn) \qquad (2)\\
&\Rightarrow\ M | q^n\\
&\Rightarrow\ M = q^l \text{with} 0 \leq l \leq n
\end{aligned}$$

In fact $l = 0$ can be ruled out because this would reduce C to a code which is trivial in the sense of only having one codeword, and the other extreme, $l = n$ would make $C = Z_q^n$ which has $d(C) = 1$ so C would not be 1-error correcting. Hence

$$M = q^l \text{ for } 0 < l < n \tag{3}$$

Returning to equation (2), this says after rearrangement that

$$n = (q^{n-l} - 1)/(q - 1) \tag{4}$$

□

From (3) $dim(C) = l$ so $dim(C^{\perp}) = n - l$. So any p.c. matrix for C will have $n - l$ rows and n columns with n given by (4). But from the proof of Theorem 6.1 this n is precisely the maximal number of columns of length $n - l$ with no pair of columns dependent. Hence, by definition C is a Hamming code.

Moving on to 2-error correcting linear codes, the condition for perfection of binary codes of dimension k is

$$M = 2^k = 2^n \Big/ \left(1 + n + \binom{n}{2}\right) = 2^{n+1}/(2 + n + n^2).$$

so $2 + n + n^2$ must be a power of 2. It was shown in 1930 that $n =$ 1, 2, 5 and 90 are the only positive integers for which this is true. Note that for 2-error correction we require $d \geq 5$ (and therefore $n \geq 5$), so the first two solutions can be ruled out. For $n = 5, M = 2$, so the code must be equivalent to a repetition code. (Exercise 12 below). We regard repetition codes as trivial, not in the sense of 'beneath contempt' – they have their uses – but because there is nothing of much interest to say about them! More sophisticated combinatorial arguments (using ideas from design theory) rule out $n = 90$. A more recent result due to Tietäväinen settles that there is no future in widening the search to $q > 3$. But for $q = 3$ there is a positive result. The perfection condition becomes $M = 3^n/(1 + 2n^2)$, so $1 + 2n^2$ must be a power of 3. $n = 11$ is a solution, and this leads to $M = 729$. So $q = 3, d = 5, n = 11, M = 729$ is a set of parameters which would give a perfect code if a code with these parameters exists. M. Golay constructed such a code, now named after him, and it is now known that any other code with these parameters is equivalent to Golay's code. Golay's code is also linear.

Work by Tietäväinen, Pless, Delsarte and Goethals up to 1975 shows just how rare perfect codes are. A summary is:

The only perfect codes with alphabet size which is prime or a power of a prime are equivalent to

(i) binary repetition codes of odd length,

(ii) Z_q^n,

(iii) all codes of the families Ham(r, q),

(iv) the Golay ternary $[11, 6, 5]$ code discussed above,

(v) the Golay binary $[23, 12, 7]$ code. [Ex 12,13]

We have previously justified calling (i) a class of trivial codes, and it is even more justifiable to dismiss (ii) in the same way since these code are 'no-error detecting' and 'no-error correcting'! (v) is still left to discuss. Before doing this the reference to prime power alphabets needs some explanation. In this book our discussions of linear codes have always assumed that the set of alphabet symbols is Z_q where q is prime. What is essential is that the alphabet should be a field and it can be shown that fields of (finite) size q exist if and only if $q = p^n$ where p is any prime and $n \geq 1$. We have effectively limited ourselves to $n = 1$. This reduces the range of useful codes we can talk about, and limits the scope of the theory accessible to us, but does not seriously prevent you from appreciating many of the fundamental ideas of the subject, which is this book's aim. The ramifications of finite field theory in coding would fill at least another book. The next sensible step if you intend using this book as a stepping stone to further coding theory would be to learn something about finite fields. The other major mathematical underpinning which does not appear in this book is design theory and finite geometry, especially if you are interested in the combinatorial aspects of coding. See the bibliography for suggestions [14], [15].

So back to item (v) in our list of perfect codes. Binary perfect codes which are 3-error correcting have $d = 7$ (and hence $n \geq 7$), and the perfection condition when simplified reduces to

$$M = 3 \times 2^{n+1}/(n+1)(n^2 - n + 6).$$

Hence $(n+1)(n^2 - n + 6)|3.2^{n+1}$. 3 and 2 are prime, so by the fundamental theorem of arithmetic one of these three cases must hold:

1. $n + 1 = 2^a, n^2 - n + 6 = 2^b$;
2. $n + 1 = 2^a.3, n^2 - n + 6 = 2^b$;
3. $n + 1 = 2^a, n^2 - n + 6 = 3.2^b$;

and in all three cases $a + b \leq n + 1$ and $n \geq 7$. Case 1 gives $(2^a - 1)^2 - (2^a - 1) + 6 = 2^b$ so

$$2^{2a} - 2^{a+1} - 2^a + 8 = 2^b \tag{1}$$

Also, for $n \geq 7, n + 1 < n^2 - n + 6$ so $a < b$ and $a \geq 3$. From (1) $2^{2a-3} - 2^{a-2} - 2^{a-3} + 1 = 2^{b-3}$. This is impossible because for $a > 3$ the left hand side is odd and the right is even, and for $a = 3$ the left is 6 which is not a power of 2. Case 2 leads by a similar argument to

$$9.2^a - 3.2^{a+1} - 3.2^a + 8 = 2b \tag{2}$$

with $a \geq 2, b \geq 6, a < b$.

Dividing (2) by 2^a, $9(2^a - 1) + \frac{8}{2^a} = 2^{b-a}$, and since 2^{b-a} is an integer, this equation cannot be satisfied unless $a = 2$ or 3.

For $a = 2$ it becomes $29 = 2^{b-2}$ which is clearly false, and for $a = 3$ it is $64 = 2^{b-3}$ so b must be 9. In this case you can check that $n = 23, d = 7$ and $M = 12^{12}$. Golay found a code with these parameters. It is also linear, and since Golay's construction of a generator matrix for it many methods have been discovered for arriving at this important code. We shall see one later in this chapter.

Case 3 leads to

$$2^a - 3 + \frac{8}{2^a} = 3.2^{b-a} \tag{3}$$

with $a \geq 3, b \geq 4, a < b$.

So the only hope of satisfying (3) is with $a = 3$, then we get $b = 4$. This gives $n = 7, M = 2, d = 7$, so we just have a repetition code.

Nothing much is known about the existence of perfect codes whose alphabet size is not a prime power.

We turn now to an upper bound specifically for linear codes. To set the scene consider (n, M, d) binary codes which by the Singleton bound must satisfy $M \leq 2^{n-d+1}$. For linear codes this becomes $2^k \leq 2^{n-d+1}$, or equivalently $k \leq n - d + 1$. We can rearrange this upper bound on k to give the equivalent lower bound on $n, n \geq k + d - 1$. Our next bound provides (usually) an improvement on this, and it involves the idea of the residual code of a linear code.

Definition 6.3 Let C be an $[n, k]$ binary code with $k \geq 2$ and let $\mathbf{c}$ be any non-zero codeword of weight $w < n$. Choose a generator matrix G whose first row is $\mathbf{c}$ with $1', 2', \cdots (n-w)'$ being the positions of its zero bits, and the rows $\boldsymbol{r}_2, \boldsymbol{r}_3, \cdots, \boldsymbol{r}_k$ are $r_{i1} r_{i2} \cdots r_{in}$ for $i = 2, 3, \cdots k$ respectively. Then the residual code $\text{Res}(C, \mathbf{c})$ is the linear code of length $n - w$ spanned by the words $r_{i1'} r_{i2'} \cdots r_{i(n-w)'}$ for $i = 2, 3, \cdots k$.

Clearly we may do a positional permutation of G to move positions $1', 2', \cdots (n-w)'$ to positions $1, 2, \cdots (n-w)$ respectively, and this has no effect on Res $(C, \mathbf{c})$. For convenience we shall always do this. [Ex 14–17]

Now we specialize a little, by choosing row 1 of G to have weight $d = d(C)$, so

$$G = \left[\begin{array}{c|c} \longleftarrow n-d \longrightarrow & \longleftarrow d \longrightarrow \\ 00\cdots0 & 11\cdots1 \\ G_1 & G_2 \end{array}\right]$$

Theorem 6.11 If C is an $[n, k, d]$ binary code with generator matrix G of the form above, then Res C has length $n - d$, dimension $k - 1$ and minimum distance $d' \geq \lceil \frac{d}{2} \rceil$. (For any real number $x \lceil x \rceil$, sometimes called the ceiling function, is the smallest integer not less than x.)

Proof. The length claim is evident from the definition. Proving that $\dim(\text{Res}C) = k - 1$ amounts to showing that the $k - 1$ rows of G_1 are independent. Suppose not. Then there is a non-trivial linear combination of them equal to $\boldsymbol{o}$. The corresponding combination of rows of G is a codeword of C, so cannot be the zero word of C. Hence it has some 1s in its last d places, but since d is the minimal non-zero weight of C it must have 1s in all these positions, which makes it identical to row 1 of G. Hence we have row 1 of G equal to a linear combination of the other rows of G, which is impossible since the rows of G are independent (being a basis of C).

Now for $d(\text{Res } C)$: let $\boldsymbol{u}$ be any non-zero word of Res C. It makes up the first $n - d$ bits of a codeword $\boldsymbol{u}|\boldsymbol{v}$ of C. [If $\boldsymbol{u} = u_1u_2 \cdots u_r$ and $\boldsymbol{v} = v_1v_2 \cdots v_s$ we use $\boldsymbol{u}|\boldsymbol{v}$ as shorthand for the word $u_1u_2 \cdots u_rv_1v_2 \cdots v_s$.] Clearly $w(\boldsymbol{u}|\boldsymbol{v}) = w(\boldsymbol{u}) + w(\boldsymbol{v})$ and since $\boldsymbol{u}|\boldsymbol{v}$ is a non-zero word of C its weight is at least d, and since it is distinct from row 1 its distance from this row is at least d.

The first of these facts implies $w(\boldsymbol{u})+w(\boldsymbol{v}) \geq d$, and the second that$w(\boldsymbol{u})+(d - w(\boldsymbol{v})) \geq d$.

Adding these we obtain $2w(\boldsymbol{u}) \geq d$, and since $w(\boldsymbol{u})$ is an integer we have $w(\boldsymbol{u}) \geq \lceil \frac{d}{2} \rceil$ as required. □

[Ex 18]

Theorem 6.12 The Griesmer bound for binary linear codes.

Let $n^*(k, d)$ denote the length of the shortest binary linear code with dimension k and minimum distance d. Then

$$n^*(k,d) \geq \sum_{i=0}^{k-1} \lceil \frac{d}{2^i} \rceil$$

Proof. Using $n(C)$ to denote the length of the code C, the result of Theorem 6.5.4 for an $[n^*(k, d), k, d]$ code C is that

$$\begin{aligned} n^*(k,d) &= d + n(\text{Res } C) && (1)\\ dim(\text{Res } C) &= k - 1 && (2)\\ d(\text{Res } C) &\geq \lceil \frac{d}{2} \rceil && (3) \end{aligned}$$

From (1) and (2) it follows that

$$n^*(k,d) \geq d + n^*(k-1, d(\text{Res } C)) \quad (4)$$

and clearly n^* is a non-decreasing function of d for fixed k, so from (3) relation (4) above implies $n^*(k,d) \geq d + n^*(k - 1, \lceil \frac{d}{2} \rceil)$, which we can apply repeatedly (with the aid of Exercise 17) to obtain

$$\begin{aligned} n^*(k,d) &\geq d + \lceil\frac{d}{2}\rceil + n^*(k-2,\lceil\frac{d}{4}\rceil) \\ &\geq d + \lceil\frac{d}{2}\rceil + \lceil\frac{d}{4}\rceil + n^*\left(k-3,\lceil\frac{d}{8}\rceil\right) \\ &\vdots \\ &\geq d + \lceil\frac{d}{2}\rceil + \lceil\frac{d}{4}\rceil + \cdots + \lceil\frac{d}{2^{k-2}}\rceil + n^*\left(1,\lceil\frac{d}{2^{k-1}}\rceil\right) \end{aligned}$$

Now it is clear that $n^*(1,d) = d$ for any d, because for a binary linear code of dimension 1 the only non-zero word must have weight d, so the shortest code possible has length d. So the last term in the sum above is $\lceil\frac{d}{2^{k-1}}\rceil$, which gives us the required result. □

For some parameters the Griesmer bound is stronger than the Hamming bound, and for cases in which the Plotkin bound is applicable it can be stronger than the Plotkin bound too. For example, for binary codes of length 20 and distance 9 the Hamming bound gives $M \leq 169$. If we are seeking a linear code with these parameters M must be a power of 2, so $M \leq 128 = 2^7$, so $k \leq 7$. But is $k = 7$ achievable? Applying the Griesmer bound we have

$$\begin{aligned} n^*(7,9) &\geq 9 + \lceil\frac{9}{2}\rceil + \lceil\frac{9}{4}\rceil + \lceil\frac{9}{8}\rceil + \lceil\frac{9}{16}\rceil + \lceil\frac{9}{32}\rceil + \lceil\frac{9}{64}\rceil \\ &= 9+5+3+2+1+1+1 \\ &= 22 \end{aligned}$$

Hence there is no binary $[20,7,9]$ code. Is there a $[20,6,9]$ binary code? We have seen that the Hamming bound does not rule this out, so we try the Griesmer bound. From the calculation above it is clear that this yields $n^*(6,9) \geq 21$ and $n^*(5,9) \geq 20$.

So, to summarize, the Hamming bound gives $k \leq 7$ but the Griesmer bound strengthens this to $k \leq 5$. [Ex 19]

The binary simplex codes provide examples of codes which are optimal by virtue of having the maximum possible length:

Theorem 6.13 All binary simplex codes meet the Griesmer bound.

Proof. Let C be the simplex code dual to $\text{Ham}(r,2)$. Then $2^r - 1, 2^{r-1}$ and r are its length, minimum distance and dimension respectively, so

$$\sum_{i=0}^{r-1}\left\lceil\frac{d}{2^i}\right\rceil = \sum_{i=0}^{r-1} 2^{r-1-i} = 2^r - 1$$

□

6.6 More on the structure of Hamming codes

There is clearly a codeword of every linear code C which is also a member of $C^{\perp}$, namely $\mathbf{0}$. If *every* codeword of C has this property, so that $C \subseteq C^{\perp}$, then C is called a *self-orthogonal* code. There are non-trivial self-orthogonal codes and the next theorems and exercises help to identify and construct some.

Theorem 6.14 The binary linear code C is self-orthogonal if and only if each generator matrix of C has all its rows of even weight and every pair of rows orthogonal.

Proof. Let G be a $k \times n$ generator matrix with the stated properties and let $\boldsymbol{c}$ be any codeword, so that

$$\boldsymbol{c} = \sum_{i=1}^{k} \lambda_i \boldsymbol{r}_i$$

where $\boldsymbol{r}_i$ is the ith row of G. Then

$$\boldsymbol{c}.\boldsymbol{r}_j = \sum_{i=1}^{k} \lambda_i(\boldsymbol{r}_i.\boldsymbol{r}_j) = \lambda_j(\boldsymbol{r}_j.\boldsymbol{r}_j) = 0$$

since $w(\boldsymbol{r})$ is even (see Exercise 9 of Chapter 5). So $\boldsymbol{c}$ is orthogonal to each row of G, and hence to every codeword of C. That is $\boldsymbol{c} \in C^{\perp}$ so $C \subseteq C^{\perp}$.

Conversely, suppose $C \subseteq C^{\perp}$. Then $\boldsymbol{r}_j.\boldsymbol{r}_j = 0$, which implies that $\boldsymbol{r}_j$ must have even weight. Secondly, if $\boldsymbol{r}_i, \boldsymbol{r}_j$ are any two rows of G then $\boldsymbol{r}_i \in C, \boldsymbol{r}_j \in C$ and hence $\boldsymbol{r}_j \in C^{\perp}$, so $\boldsymbol{r}_i.\boldsymbol{r}_j = 0$. □

[Ex 20]

The previous theorem has the following natural variation for codes over Z_3.

Theorem 6.15 The ternary code C is self-orthogonal if and only if each generator matrix of C has all its row-weights equal to a multiple of 3 and every pair of rows orthogonal.

Proof. Let G be a generator matrix with the stated properties and let $\boldsymbol{r}$ be any row of G. Now $\boldsymbol{r}.\boldsymbol{r}$ is a sum of $w(\boldsymbol{r})$ non-zero terms each of which is 1 since $1^2 \equiv 2^2 \equiv 1 \bmod 3$. So $\boldsymbol{r}.\boldsymbol{r} = 0$ if and only if $w(\boldsymbol{r}) \equiv 0 \bmod 3$. The rest of the proof is identical to the binary case. □

[Ex 21]

In spite of Exercise 21 it is possible to generalize these theorems to cover all p by the somewhat weaker result:

Theorem 6.16 The linear code C over Z_p is self-orthogonal if and only if each generator matrix has all its rows satisfying $\boldsymbol{r}_i.\boldsymbol{r}_i = 0$ and $\boldsymbol{r}_i.\boldsymbol{r}_j = 0$.

Proof. Obvious from the proof of Theorem 6.14. □

The result of Exercise 20 can also be generalized to $p = 3$:

Theorem 6.17 All ternary simplex codes of dimension $k \geq 2$ are self-orthogonal.

Proof. Let T_k be the matrix whose columns are all the ternary strings of length k and let S_k be a parity check matrix of the code $C_k \in \text{Ham}(k, 3)$ defined by the construction of section 6.2. Then S_k is a generator matrix for a ternary simplex code of dimension k. S_k has the form:

$$\left[\begin{array}{c|c} 1\,1\cdots 1 & 00\cdots 0 \\ & \\ T_{k-1} & S_{k-1} \end{array}\right] \tag{1}$$

in which row 1 consists of 3^{k-1} ones followed by $\frac{1}{2}\left(3^{k-1}-1\right)$ zeros. By the proof of Theorem 6.6 all rows of S_k have weight 3^{k-1} which is a multiple of 3 since $k \geq 2$.

Now we show that the rows of S_k satisfy the other condition of Theorem 6.15 and we do this by induction on k.

S_2 is $\begin{bmatrix} 1 & 1 & 1 & 0 \\ 0 & 1 & 2 & 1 \end{bmatrix}$, from which it is easy to check that (row 1. row 2) $= 0$. Now assume that it holds for all dimensions from 2 to $k-1 (k \geq 3)$, and consider S_k. Note that by symmetry each row of T_{k-1} consists of 3^{k-2} occurrences of each of 0, 1 and 2, (2)
and that each pair of distinct rows of T_{k-1} contains 3^{k-3} occurrences of each of the nine symbol pairs

$$\begin{matrix} 0 & 0 & 0 & 1 & 1 & 1 & 2 & 2 & 2 \\ 0, & 1, & 2, & 0, & 1, & 2, & 0, & 1, & 2. \end{matrix} \tag{3}$$

From (1)

$$\begin{aligned} &(\text{row } 1.\ \text{row } i) && (i \neq 1) \\ &\equiv \text{sum of symbols in row } i \text{ of } T_{k-1} \\ &= \ 3^{k-2}(1+2) && \text{by (2)} \\ &\equiv 0 \bmod 3 \end{aligned}$$

and from (1) $(\text{row } i.\ \text{row } j)$ $(1 < i < j)$

$$\begin{aligned} &= (\text{row } i.\ \text{row } j) \text{ of } T_{k-1} + (\text{row } i.\ \text{row } j) \text{ of } S_{k-1} \\ &= 3^{k-3}(1.1 + 1.2 + 2.1 + 2.2) + 0 \\ &\quad \text{by (3) and the induction hypothesis} \\ &= 3^{k-1} \\ &\equiv 0 \bmod 3. \end{aligned}$$

This completes the inductive proof that ternary simplex codes *with gen-*

erator matrices of the form (1) are self-orthogonal, but there are of course other ternary simplex codes – those with generator matrices whose columns do not all have 1 as their first non-zero entries. This hole is easy to plug since, by the construction given in the proof of Theorem 6.1, any of these more general generator matrices will be the result of taking one of the simpler matrices and multiplying some of its columns by 2. This will not change the weight of any row, nor will it change the dot product of any two rows since each pair of symbols in Z_3, when doubled, becomes a pair with the same product modulo 3. □

The idea at the end of the previous proof gives a hint that there is another respect in which Z_2 and Z_3 are rather special alphabets:

Theorem 6.18 Self-orthogonality for codes over Z_2 and Z_3 is an equivalence-invariant property. That is, if C is a self-orthogonal binary or ternary linear code and C is equivalent to C', then C' is self-orthogonal.

Proof. Let G be any generator matrix for C. Then G has the properties given by Theorem 6.16, and doing row operations $R1, R2, R3$ and column operation $C1$ on G clearly will not change this. Column operation $C2$ does nothing in the binary case, and for ternary codes will only double some of the columns. In this case the remark above makes it clear that the orthogonality of the rows is preserved. □

[Ex 22]

If we have a self-orthogonal $[n, k]$ code C whose dimension is half its length then we have $C \subseteq C^{\perp}$ and $\dim C = \dim C^{\perp} = \frac{n}{2}$, from which it follows (item 11(b) of section 5.3) that $C = C^{\perp}$. Such codes are called self-dual codes. Clearly no *binary* Hamming code can be self-dual because their lengths are odd. [Ex 23]

If however we extend the 7-bit binary Hamming code by adding an overall parity check bit as in Chapter 2, then the extended code C has p.c. matrix

$$\begin{bmatrix} 1 & 0 & 1 & 1 & 1 & 0 & 0 & 0 \\ 1 & 1 & 0 & 1 & 0 & 1 & 0 & 0 \\ 1 & 1 & 1 & 0 & 0 & 0 & 1 & 0 \\ 1 & 1 & 1 & 1 & 1 & 1 & 1 & 1 \end{bmatrix}.$$

This is a generator matrix for $C^{\perp}$ so $C^{\perp}$ has length 8 and dimension 4, and it is easy to check that this matrix satisfies the conditions to generate a self-orthogonal code. Hence $C^{\perp}$ is self-dual so $C^{\perp} = (C^{\perp})^{\perp} = C$.

[Ex 24, 25]

Unfortunately this result does not extend to the larger binary Hamming codes. But their duals are important codes : they provide one way of defining the first order Reed–Muller codes, and in Chapter 9 we discuss two further methods of constructing them.

6.7 The cyclic property of Hamming codes

We shall have much more to say about cyclic codes in Chapter 8, but for the moment we just draw your attention to a nice cyclic symmetry of some Hamming codes. This cyclic property is then used to construct the binary $[23, 12, 7]$ Golay code. The apparently arbitrary first step will appear in a more natural light in Chapter 8. For the method I am indebted to Pretzel [16].

Consider the code C in Ham(3, 2) given by the p.c. matrix

$$H = \begin{bmatrix} 1 & 0 & 1 & 1 & 1 & 0 & 0 \\ 0 & 1 & 0 & 1 & 1 & 1 & 0 \\ 0 & 0 & 1 & 0 & 1 & 1 & 1 \end{bmatrix}$$

From this we obtain a complete list of the sixteen codewords displayed below. Alongside each word of C we have written the word obtained by writing its bits in reverse order, and the code D is this set of reversals.

	C	D
P	0 0 0 0 0 0 0	0 0 0 0 0 0 0
	1 1 0 1 0 0 0	0 0 0 1 0 1 1
	0 1 1 0 1 0 0	0 0 1 0 1 1 0
	0 0 1 1 0 1 0	0 1 0 1 1 0 0
Q	0 0 0 1 1 0 1	1 0 1 1 0 0 0
	1 0 0 0 1 1 0	0 1 1 0 0 0 1
	0 1 0 0 0 1 1	1 1 0 0 0 1 0
	1 0 1 0 0 0 1	1 0 0 0 1 0 1
	0 0 1 0 1 1 1	1 1 1 0 1 0 0
	1 0 0 1 0 1 1	1 1 0 1 0 0 1
	1 1 0 0 1 0 1	1 0 1 0 0 1 1
R	1 1 1 0 0 1 0	0 1 0 0 1 1 1
	0 1 1 1 0 0 1	1 0 0 1 1 1 0
	1 0 1 1 1 0 0	0 0 1 1 1 0 1
	0 1 0 1 1 1 0	0 1 1 1 0 1 0
S	1 1 1 1 1 1 1	1 1 1 1 1 1 1

The sixteen codewords of C and D have been written in blocks P, Q, R, S containing respectively the words of weights 0, 3, 4 and 7. The order of the words in blocks Q and R of C have the property that each word is the result of moving the right hand bit of the previous word to the left hand end, so D also has this cyclic property but in the opposite direction. Another symmetry of both codes is that each word of R is the complement of the corresponding word in Q. Finally, codes C and D only have the words of weights 0 and 7 in common. [Ex 26]

This implies that C (and hence D) are *cyclic* codes. That is, for each codeword $a_1a_2\cdots a_n$, $a_na_1a_2\cdots a_{n-1}$ is also a codeword. Also C and D have the property that the complement of each codeword is a codeword.

If we now extend both C and D to C', D', by adding an overall parity check bit to each word, thus making every word of even weight, we shall have two linear codes whose weights are all 0, 4 or 8, having only the words of weight 0 and 8 in common.

Using C' and D' we construct words of length 24 as follows: the first 8 bits are $\boldsymbol{a}+\boldsymbol{x}$ where $\boldsymbol{a}\in C'$, $\boldsymbol{x}\in D'$; the next 8 are $\boldsymbol{b}+\boldsymbol{x}$ where $\boldsymbol{b}\in C'$; and the last 8 are $\boldsymbol{a}+\boldsymbol{b}+\boldsymbol{x}$. Let the code E' be the set of all such words, so in the notation introduced at the end of section 4.6,

$$E' = \{\boldsymbol{a}+\boldsymbol{x}|\boldsymbol{b}+\boldsymbol{x}|\boldsymbol{a}+\boldsymbol{b}+\boldsymbol{x} \ : \boldsymbol{a}\in C', \boldsymbol{b}\in C', \boldsymbol{x}\in D'\}$$

[Ex 27]

So E' is a linear code of dimension 12. The next exercises set up what we need to prove E' has a minimum distance of 8. [Ex 28, 29]

From the form given for E' we can write any of its codewords as the sum $\boldsymbol{e}=\boldsymbol{a}|\boldsymbol{0}|\boldsymbol{a}+\boldsymbol{0}|\boldsymbol{b}|\boldsymbol{b}+\boldsymbol{x}|\boldsymbol{x}|\boldsymbol{x}$, so by the result of Exercise 28,

$$\begin{aligned} w(\boldsymbol{e}) &= w(\boldsymbol{a}|\boldsymbol{0}|\boldsymbol{a})+w(\boldsymbol{0}|\boldsymbol{b}|\boldsymbol{b})+w(\boldsymbol{x}|\boldsymbol{x}|\boldsymbol{x})-2w((\boldsymbol{a}|\boldsymbol{0}|\boldsymbol{a})\odot(\boldsymbol{0}|\boldsymbol{b}|\boldsymbol{b}))\\ &\quad -2w((\boldsymbol{a}|\boldsymbol{0}|\boldsymbol{a})\odot(\boldsymbol{x}|\boldsymbol{x}|\boldsymbol{x}))-2w((\boldsymbol{0}|\boldsymbol{b}|\boldsymbol{b})\odot(\boldsymbol{x}|\boldsymbol{x}|\boldsymbol{x}))\\ &\quad +4w(((\boldsymbol{a}|\boldsymbol{0}|\boldsymbol{a})\odot(\boldsymbol{x}|\boldsymbol{x}|\boldsymbol{x}))\odot((\boldsymbol{0}|\boldsymbol{b}|\boldsymbol{b})\odot(\boldsymbol{x}|\boldsymbol{x}|\boldsymbol{x})))\\ &= 2w(\boldsymbol{a})+2w(\boldsymbol{b})+3w(\boldsymbol{x})-2w(\boldsymbol{a}\odot\boldsymbol{b})-4w(\boldsymbol{a}\odot\boldsymbol{x})-4w(\boldsymbol{b}\odot\boldsymbol{x})\\ &\quad +4w((\boldsymbol{a}\odot\boldsymbol{x}|\boldsymbol{0}|\boldsymbol{a}\odot\boldsymbol{x})\odot(\boldsymbol{0}|\boldsymbol{b}\odot\boldsymbol{x}|\boldsymbol{b}\odot\boldsymbol{x})) \end{aligned}$$

Now $\boldsymbol{a}$, $\boldsymbol{b}$, $\boldsymbol{x}$ all have weights divisible by 4, and by Exercise 29 $w(\boldsymbol{a}\odot\boldsymbol{b})$ is even, so all terms in the sum above are multiples of 4. There is a word of weight 8 in E' – for example take $\boldsymbol{a}=\boldsymbol{x}=11111111$ and $\boldsymbol{b}=11010001$, so it just remains to show there is no word of weight 4. Aiming for a contradiction, suppose $\boldsymbol{e}=\boldsymbol{a}+\boldsymbol{x}|\boldsymbol{b}+\boldsymbol{x}|\boldsymbol{a}+\boldsymbol{b}+\boldsymbol{x}$ is such a word. Now $\boldsymbol{a}$, $\boldsymbol{b}, \boldsymbol{x}$ all have even weight so it follows from the '$w(\boldsymbol{x}+\boldsymbol{y})$' formula that $\boldsymbol{a}+\boldsymbol{x}, \boldsymbol{b}+\boldsymbol{x}$ and $\boldsymbol{a}+\boldsymbol{b}+\boldsymbol{x}$ are all even weight. Since their total weight is supposedly 4, at least one of them has zero weight, so $\boldsymbol{x}=\boldsymbol{a}$ or $\boldsymbol{b}$ or $\boldsymbol{a}+\boldsymbol{b}$. That is $\boldsymbol{x}\in C'$, so by linearity all of $\boldsymbol{a}+\boldsymbol{x}$, $\boldsymbol{b}+\boldsymbol{x}, \boldsymbol{a}+\boldsymbol{b}+\boldsymbol{x}$ are in C', so they all have weight zero or at least 4. Hence exactly two of them have weight 0 and the other has weight 4. This implies that exactly two of $\boldsymbol{a}$, $\boldsymbol{b}, \boldsymbol{a}+\boldsymbol{b}$ are equal to $\boldsymbol{x}$ and the third is $\boldsymbol{0}$. We have shown $\boldsymbol{x}\in C'\cap D'$ so $\boldsymbol{x}$ is $\boldsymbol{0}$ or 11111111, and $\boldsymbol{e}=\boldsymbol{0}|\boldsymbol{0}|\boldsymbol{x}$ or $\boldsymbol{0}|\boldsymbol{x}|\boldsymbol{0}$ or $\boldsymbol{x}|\boldsymbol{0}|\boldsymbol{0}$, all of which have weight 0 or 8, so the proof is complete.

Since E' is an even weight code (with $d=8$) it can be regarded as an extension of a code E of length 23. E has minimum weight 7 since we have already seen an example of a codeword of E' with 1 as its last bit. So E has the parameters of the binary Golay code, and by the remark in section 6.5, E is equivalent to this code.

There is a neat connection between cyclicity and duality:

Theorem 6.19 if C is cyclic then so is $C^{\perp}$.

Proof. Let C be a cyclic $[n, k]$ code. Let $\boldsymbol{c}$ be any codeword and $\boldsymbol{c}^t$ its tth cyclic shift; that is, if $\boldsymbol{c} = c_1\, c_2 \cdots c_n$ then $\boldsymbol{c}^t = c_{n-t+1}c_{n-t+2}\cdots c_n c_1 c_2 \cdots c_{n-t}$. Note that is is immediate from the definition of a cyclic code that if $\boldsymbol{c} \in C$, then $\boldsymbol{c}^t \in C$ for all t. We show that $\boldsymbol{h}^1 \in C^{\perp}$ whenever $\boldsymbol{h} \in C^{\perp}$ thus proving the cyclicity of $C^{\perp}$.

$$\begin{aligned}
\boldsymbol{h}^1 \cdot \boldsymbol{c} &= h_n c_1 + h_1 c_2 + \cdots + h_{n-1} c_n \\
&= h_1 c_2 + \cdots + h_{n-1} c_n + h_n c_1 \\
&= \boldsymbol{h} \cdot \boldsymbol{c}^{n-1} = \boldsymbol{0} \text{ because } \boldsymbol{h} \in C^{\perp} \text{ and } \boldsymbol{c}^{n-1} \in C \\
\text{Hence} \quad & \boldsymbol{h}^1 \in C^{\perp} \text{ as required.}
\end{aligned}$$

□

Unfortunately cyclicity is not equivalence invariant. (Note that we had to exercise some care in choosing a member of Ham(3,2) which is cyclic.) So the question arises: for which r and q does Ham(r, q) have a cyclic representation? The question is important because cyclic codes have hardware and software implementation advantages over non-cyclic codes. [Ex 30]

The answer is that Ham$(r, 2)$ has a cyclic representative for all r, as do the non-binary ones with $gcd(r, q-1) = 1$, but the proof of this requires algebraic machinery beyond the scope of this book. Combining the result with Theorem 6.19 we obtain the corollary that all binary simplex codes have cyclic representatives.

6.8 Weight distributions

We have mentioned several times that minimum distance is not a particularly precise measure of a code's overall performance in processing errors. To illustrate, consider the case of a linear code C being used for error detection, and ask what is the probability that we fail to detect an error. Let the alphabet be Z_q and the length n. The required probability is that of the received word being a codeword yet containing at least one error. It is easy to see, using the linearity of C, that this is equivalent to the error vector $\boldsymbol{e}$ being a non-zero codeword. Now there is probability

$$\binom{n}{i} p^i (1-p)^{n-i}$$

that the error vector has weight i where p is the probability that an arbitrary bit is corrupted. There are

$$\binom{n}{i}(q-1)^i$$

words of weight i, and if A_i is the number of codewords of weight i, then the probability that $\mathbf{e}$ is a codeword of weight i is

$$A_i\Big/\binom{n}{i}(q-1)^i.$$

Hence

$$\begin{aligned}\text{Prob }[w(\mathbf{e})=i] &= \frac{A_i}{\binom{n}{i}(q-1)^i}\times\binom{n}{i}p^i(1-p)^{n-i}\\ &= A_i p^i(1-p)^{n-i}(q-1)^{-i},\end{aligned}$$

so

$$\text{Prob }[\mathbf{e}\text{ is a non-zero codeword }]=\sum_{i=1}^{n}A_i p^i(1-p)^{n-i}(q-1)^{-i}.$$

A good error-detecting code will be one for which this probability is small. For small p (the usual situation in practice) $p^i(1-p)^{n-i}$ rapidly decreases as i increases, so we would like A_i to be small when i is. That is, C should ideally have few codewords of low weight and most of them with large weight. The weight distribution cannot of course be chosen to order – the linear structure imposes severe constraints on it. But as we have seen, the calculation of the code's performance depends on knowledge of the A_is, and it turns out that many theoretical investigations of linear codes are dependent on their weight distributions.

One of the most famous aids to progress is the MacWilliams identity, which is a remarkable connection between the weight distributions of C and $C^\perp$. Its significance is that the weight distribution of $C^\perp$ is often simpler to calculate than that of C, and the identity enables each distribution to be obtained from the other. Since its discovery in 1963 it has been expressed in many different forms and has been generalized in various ways. We end this chapter with a proof of one of its forms and a couple of simple examples of its use. The proof we give makes minimal use of complex numbers.

Let C be a linear $[n,k]$ code over Z_q for some prime q. Let χ be the function which maps each α in Z_q to the complex number $\exp(2\pi i\alpha/q)$, so in the complex plane the χ-images of Z_q are q points equally spaced round the unit circle with

$$\chi(o)=1 \qquad (1)$$

and, by the properties of multiplication of complex numbers, for all α,β in

Z_q we have

$$\chi(\alpha + \beta) = \chi(\alpha)\chi(\beta) \qquad (2)$$

We require two preliminary lemmas:

Lemma 6.1 $\sum_{\alpha \in Z_q} \chi(\alpha) = 0$

Proof. Let β be any non-zero member of Z_q so that $\chi(\beta) \neq 1$. If α_1 and α_2 are distinct members of Z_q, then so are $\alpha_1 + \beta$ and $\alpha_2 + \beta$, so if we choose a fixed $\beta \neq 0$ and let α take each value in Z_q in turn then $\alpha + \beta$ takes each value in Z_q once and once only. Hence

$$\begin{aligned}\sum_{\alpha \in Z_q} \chi(\alpha) &= \sum_{\alpha \in Z_q} \chi(\alpha + \beta) \\ &= \sum_{\alpha \in Z_q} \chi(\alpha)\chi(\beta) \text{ from (2)} \\ &= \chi(\beta) \sum_{\alpha \in Z_q} \chi(\alpha) \text{ since } \beta \text{ is fixed.}\end{aligned}$$

So

$$\left[\sum_{\alpha \in Z_q} \chi(\alpha)\right] [1 - \chi(\beta)] = 0, \text{ and since the complex numbers}$$

form a field, one of these two factors must be zero, but $1 - \chi(\beta) \neq 0$ since $\beta \neq 0$, so $\chi(\beta) \neq 1$.

Hence $\sum_{\alpha \in Z_q} \chi(\alpha) = 0$ □

Next we need another symmetry property of linear codes.

Lemma 6.2 Let $\boldsymbol{w}$ be any fixed member of Z_q^n not in $C^\perp$. For each $\alpha \in Z_q$ define the subset S_α of C by $S_\alpha = \{\boldsymbol{c} \in C : \boldsymbol{c} \cdot \boldsymbol{w} = \alpha\}$

Then all these subsets have the same size, which is non-zero.

Proof. S_0 is non-empty because it is easy to check from the definition of S_0 that $\mathbf{0} \in S_0$. Also, S_0 is a linear code because, again from the definition of S_0,

$$\boldsymbol{c} \in S_0 \Rightarrow \boldsymbol{c}_1 \cdot \boldsymbol{w} = 0 \Rightarrow (\lambda \boldsymbol{c}) \cdot \boldsymbol{w} = 0 \text{ for all } \lambda \in Z_q$$

$$\text{and } \boldsymbol{c}_1, \boldsymbol{c}_2 \in S_0 \Rightarrow \boldsymbol{c}_1 \cdot \boldsymbol{w} = \boldsymbol{c}_2 \cdot \boldsymbol{w} = 0 \Rightarrow (\boldsymbol{c}_1 + \boldsymbol{c}_2) \cdot \boldsymbol{w} = 0.$$

So now we consider cosets of the linear code S_0, and we show that provided S_α is not empty and $\mathbf{c}'$ is any of its members, then S_α is in fact the coset $\mathbf{c}' + S_0$:

$\mathbf{c}' \in S_\alpha \Rightarrow \mathbf{c}' \in \mathbf{c}' + S_0$ because $\mathbf{c}' = \mathbf{c}' + \mathbf{0}$ and $\mathbf{0} \in S_0$, and conversely, $\mathbf{c}'' \in \mathbf{c}' + S_0 \Rightarrow \mathbf{c}'' = \mathbf{c}' + \mathbf{c}$ for some $\mathbf{c} \in S_0$

$$\Rightarrow \mathbf{c}'' \cdot \boldsymbol{w} = \mathbf{c}' \cdot \boldsymbol{w} + \mathbf{c} \cdot \boldsymbol{w} = \alpha + 0 = \alpha, \text{ so } \mathbf{c}'' \in S_\alpha.$$

This completes the proof that each non-empty S_α is a coset of S_0, so they are all the same size. It just remains to show that there is no α for which S_α is empty.

Let α be any non-zero member of Z_q. $\boldsymbol{w} \notin C^\perp$ so there exists $\mathbf{c} \in C$ such that $\mathbf{c} \cdot \boldsymbol{w} = \lambda \neq 0$. Consider the set of codewords $\{\theta\mathbf{c} : \theta \in Z_q\}$, and their dot products with $\boldsymbol{w}$:

$$(\theta\mathbf{c}) \cdot \boldsymbol{w} = \theta(\mathbf{c} \cdot \boldsymbol{w}) = \theta\lambda,$$

But $\{\theta\lambda : \theta \in Z_q\} = Z_q$ (see proof of Fermat's theorem in Chapter 3).

So there is some θ for which $\theta\lambda = \alpha$, so for this $\theta, (\theta\mathbf{c}) \cdot \boldsymbol{w} = \alpha$. Hence $\theta\mathbf{c} \in S_\alpha$ so $S_\alpha \neq \phi$. □

We are now ready to prove a version of the MacWilliams identity. It will be convenient to specify the weight distribution of C by a formal polynomial.

$A(x) = A_0 + A_1x + A_2x^2 + \cdots + A_nx^n$, called the weight enumerator of C, and the A_i are as previously defined.

Theorem 6.20 (The MacWilliams identity) *Let C be an $[n,k]$ linear code over Z_q and $A(x), B(x)$ the weight enumerators of $C, C^\perp$ respectively.*

$$B(x) \stackrel{(1)}{=} |C|^{-1}(1+(q-1)x)^n A\left(\frac{1-x}{1+(q-1)x}\right)$$

$$\stackrel{(2)}{=} |C|^{-1}\sum_{i=0}^{n} A_i(1-x)^i(1+(q-1)x)^{n-i}$$

Proof. The easy bit (equality (2) above) is covered by the next exercise. [Ex 31]

Now we prove (1). With χ as in Lemma 6.1 let

$$J = \sum_{\mathbf{c}\in C}\left[\sum_{\boldsymbol{u}\in Z_q^n} \chi(\mathbf{c}\cdot\boldsymbol{u})x^{w(\boldsymbol{u})}\right] \tag{3}$$

If we think of this as the sum

$$\sum_{\boldsymbol{u}\in Z_q^n} \mu_{\boldsymbol{u}} x^{w(\boldsymbol{u})},$$

the coefficient $\mu_{\boldsymbol{u}}$ is

$$\sum_{\boldsymbol{c}\in C}\chi(\boldsymbol{c}\cdot\boldsymbol{u}),$$

so (3) can be written as

$$J=\sum_{\boldsymbol{u}\in Z_q^n}\left[x^{w(\boldsymbol{u})}\sum_{\boldsymbol{c}\in C}\chi(\boldsymbol{c}\cdot\boldsymbol{u})\right] \tag{4}$$

In this sum consider those terms arising from the words $\boldsymbol{u}$ in $C^\perp$ and from those not in $C^\perp$ separately.

When $\boldsymbol{u}\in C^\perp$

$$\sum_{\boldsymbol{c}\in C}\chi(\boldsymbol{c}\cdot\boldsymbol{u})=\sum_{\boldsymbol{c}\in C}\chi(0)=\sum_{\boldsymbol{c}\in C}1=|C| \tag{5}$$

and for each fixed $\boldsymbol{u}$ not in $C^\perp$ Lemma 6.2 tells us that $\boldsymbol{c}\cdot\boldsymbol{u}$ takes each value in Z_q exactly q^{k-1} times as $\boldsymbol{c}$ ranges over C. Hence, for these $\boldsymbol{u}$,

$$\sum_{\boldsymbol{c}\in C}\chi(\boldsymbol{c}\cdot\boldsymbol{u})=q^{k-1}\sum_{\alpha\in Z_q}\chi(\alpha)=0,\ \text{by Lemma 6.1.} \tag{6}$$

From (5) and (6) we see that only those $\boldsymbol{u}$ in $C^\perp$ contribute anything to the sum (4), so it can now be rewritten as

$$J=\sum_{\boldsymbol{u}\in C^\perp}x^{w(\boldsymbol{u})}|C|=|C|\sum_{\boldsymbol{u}\in C^\perp}x^{w(\boldsymbol{u})}$$

Now $C^\perp$ has B_i words of weight i, so grouping all words of the same weight together we finally get

$$J=|C|\sum_{i=1}^{n}B_ix^i=|C|B(x) \tag{7}$$

Now we manipulate the expression on the right of (3) in a different way, by expanding its inner sum. Let $\boldsymbol{u}$ be the word $u_1u_2\cdots u_n$. We shall abuse notation slightly and write $w(u_i)=0$ or 1 when u_i is zero or non-zero respectively, so that $w(\boldsymbol{u})=\sum_{i=1}^{n}w(u_i)$. Then

$$\sum_{\boldsymbol{u}\in Z_q^n}\chi(\boldsymbol{c}\cdot\boldsymbol{u})x^{w(\boldsymbol{u})}=\sum_{u_i\in Z_q}\chi(c_1u_1+\cdots+c_nu_n)x^{w(u_1)+\cdots+w(u_n)},$$

which, by property (1) is

$$\sum_{u_i\in Z_q}\chi(c_1u_1)\cdots\chi(c_nu_n)x^{w(u_1)}\cdots x^{w(u_n)} \tag{$*$}$$

The q^n terms of this sum are precisely what we would get by taking, for each i, the sum

$$\chi(c_i \times 0)x^{w(0)} + \chi(c_i \times 1)x^{w(1)} + \chi(c_i \times 2)x^{w(2)} + \cdots + \chi(c_i \times (q-1))x^{w(q-1)}$$

and multiplying them all together. So $(*)$ can be written as

$$\sum_{\boldsymbol{u} \in Z_q^n} \chi(\boldsymbol{c} \cdot \boldsymbol{u})x^{w(\boldsymbol{u})} = \prod_{i=1}^{n} \sum_{u_i \in Z_q} \chi(c_i u_i)x^{w(u_i)}. \tag{8}$$

Now look at the summation in (8). That is, fix i and let u_i range over Z_q.

If $c_i = 0$ then

$$\begin{aligned} \sum_{u_i \in Z_q} \chi(c_i u_i)x^{w(u_i)} &= \sum_{u_i \in Z_q} \chi(0)x^{w(u_i)} \\ &= \sum_{u_i \in Z_q} x^{w(u_i)} \text{ from (1)} \\ &= 1 + (q-1)x \qquad (9) \end{aligned}$$

since $w(0) = 0$ and $w(u_i) = 1$ for the other $q-1$ members of Z_q. If $c_i \neq 0$, then by splitting off the $u_i = 0$ term from the rest we get

$$\sum_{u_i \in Z_q} \chi(c_i u_i)x^{w(u_i)} = \chi(0)x^0 + \sum_{u_i \neq 0} \chi(c_i u_i)x^{w(u_i)}$$

$$= 1 + \sum_{u_i \neq 0} \chi(c_i u_i)x = 1 + \sum_{u_i \neq 0} \chi(u_i)x,$$

(as in the proof of Fermat's theorem again)

$$= 1 + (-x) \text{ from Lemma 6.1.} \tag{10}$$

Use (9) and (10) in (8) we get

$$\sum_{\boldsymbol{u} \in Z_q^n} \chi(\mathbf{c} \cdot \boldsymbol{u})x^{w(\boldsymbol{u})} = \prod_{i=1}^{n} \sum_{u_i \in Z_q} \chi(c_i u_i)x^{w(u_i)} = (1+(q-1)x)^{n-w(\mathbf{c})}(1-x)^{w(\mathbf{c})} \tag{11}$$

since the first factor is the contribution of those is ($n - w(\mathbf{c})$ of them) for which $c_i = 0$ and the second factor comes from the remaining $w(\mathbf{c})$$i$s.

So from (3)

$$\begin{aligned} J &= \sum_{\mathbf{c} \in C} (1 + (q-1)x)^{n-w(\mathbf{c})} (1-x)^{w(\mathbf{c})} \\ &= \sum_{i=0}^{n} A_i (1 + (q-1)x)^{n-i}(1-x)^i \qquad (12) \end{aligned}$$

and by equating the expressions for J in (12) and (7) the equality (1) follows. □

To see an easy application of the MacWilliams identity consider the simplest of all linear codes, the repetition codes. The repetition code C of length n over Z_q has generator $G = [11 \cdots 1]$ and its weight enumerator is $A(x) = 1 + (q-1)x^n$. Hence its dual code has weight enumerator

$$B(x) = \frac{1}{q}\left[(1+(q-1)x)^n + (q-1)(1-x)^n\right]$$

[Ex 32, 33]

Now let us investigate the weight distributions of Hamming codes. Let $C \in \text{Ham}(r,q)$. We know that all members of $\text{Ham}(r,q)$ are equivalent so their distance distributions are the same. Since the codes are linear this implies their weight distributions are the same (see Theorem 6.5).

We also know (Theorem 6.6) that the simplex code $C^\perp$ has one word of weight 0 and all the remaining $q^r - 1$ codewords have weight q^{r-1}. Hence $B_0 = 1, B_{q^{r-1}} = q^r - 1$ and all the other B_i are zero. From the MacWilliams relation, interchanging $B(x)$ and $A(x)$ (which is valid because $(C^\perp)^\perp = C$), we have

$$A(x) = \frac{1}{q^r}\Big[B_0(1-x)^0(1+(q-1)x)^n + B_{q^{r-1}}(1-x)^{q^{r-1}}(1+(q-1)x)^{n-q^{r-1}}\Big]$$

and using $n = \dfrac{q^r-1}{q-1}$ this simplifies to

$$A(x) = \frac{1}{q^r}\left[(1+(q-1)x)^n + (q^r-1)(1-x)^{q^{r-1}}(1+(q-1)x)^{\frac{q^{r-1}}{q-1}}\right]$$

[Ex 34]

As an alternative to working out the coefficients of powers of x in the expression above we can make use of the fact that Hamming codes are *perfect* one-error correcting codes. This implies that, given any word in Z_q^n, that word will either be a codeword, or it will be at distance 1 from a unique codeword. Hence we can count the total number of words of weight w, $\binom{n}{w}(q-1)^w$, by adding $X+Y+Z$ where X is the number of codewords of weight w, Y is the number of words of weight w at distance 1 from a codeword of weight $w+1$ and Z is the number of words of weight w at distance 1 from a codeword of weight $w-1$.

$$\begin{aligned} X &= A_w \\ Y &= A_{w+1}(w+1) \\ Z &= A_{w-1}(n-w+1)(q-1) \end{aligned}$$

[Ex 35]

Hence

$$\binom{n}{w}(q-1)^w = A_w + A_{w+1}(w+1) + A_{w-1}(n-w+1)(q-1).$$

Then if any two consecutive A_i are known, all subsequent ones can be calculated. [Ex 36]

We have used the weight distribution numbers to calculate how well a linear code performs with regard to error detection. They also yield an estimate of the error correction performance, as we now demonstrate. We concentrate on linear binary codes. Linearity makes life simpler because the probability of the received word being further from the transmitted word than from some other codeword is independent of the transmitted word (this follows from Theorem 5.3). So let $\mathbf{c} \in C$, a linear binary code of length n, with $\mathbf{c} \neq \mathbf{0}$, and suppose $\mathbf{0}$ is the transmitted word. As usual we take the decoding scheme to be nearest neighbour, and we assume that if there is more than one candidate for the decoded word we request retransmission. Hence there is a decoding error or a retransmission request if $d(\boldsymbol{r}, \mathbf{0}) \geq d(\boldsymbol{r}, \mathbf{c})$. So we first calculate the probability $P_{\mathbf{c}}$ that this relation holds for a fixed $\mathbf{c} \neq \mathbf{0}$. If $\mathbf{c}$ has weight w let there be x positions in which $\mathbf{c}$ and $\boldsymbol{r}$ both have a 1 and y positions in which $\mathbf{c}$ has a 0 and $\boldsymbol{r}$ has 1. Then $d(\boldsymbol{r}, \mathbf{c}) = w - x + y$ and $d(\boldsymbol{r}, \mathbf{0}) = x + y$.

Hence

$$\begin{aligned} P_{\mathbf{c}} &= P[x + y \geq w - x + y] = P[x \geq \frac{w}{2}] \\ &= P[x \geq \lceil \frac{w}{2} \rceil] \text{ since } x \text{ is an integer} \\ &= \sum_{i=\lceil \frac{w}{2} \rceil}^{w} \binom{w}{i} p^i (1-p)^{w-i} \text{ where } p \text{ is the probability} \end{aligned}$$

that the channel corrupts any given bit. To ease the notation we write a' for $\lceil \frac{a}{2} \rceil$. If $P_{\mathbf{c}}$ is now summed over all non-zero codewords $\mathbf{c}$, the result is an upper bound for the probability $P(\text{error})$ that $\boldsymbol{r}$ is as close as or is closer to at least one non-zero codeword than it is to $\mathbf{0}$ – in other words, that a decoding error or a retransmission request is made. Carrying out this summation we obtain

$$\begin{aligned} P(\text{error}) &= \sum_{\mathbf{c} \neq \mathbf{0}} \sum_{i=w'}^{w} \binom{w}{i} p^i (1-p)^{w-i} \\ &= \sum_{j=1}^{n} A_j \sum_{i=j'}^{j} \binom{j}{i} p^i (1-p)^{j-i} \end{aligned}$$

$$\begin{aligned}
&\le \sum_{j=1}^{n} A_j \sum_{i=j'}^{j} \binom{j}{i} p^{j'}(1-p)^{j-j'} && * \\
&\le \sum_{j=1}^{n} A_j \sum_{i=j'}^{j} \binom{j}{i} p^{\frac{j}{2}}(1-p)^{\frac{j}{2}} && * \; * \\
&= \sum_{j=1}^{n} A_j p^{\frac{j}{2}}(1-p)^{\frac{j}{2}} \sum_{i=j'}^{j} \binom{j}{i} \\
&\le \sum_{j=1}^{n} A_j \sqrt{p^j(1-p)^j}\, 2^j && \text{since} \sum_{i=0}^{j} \binom{j}{i} = 2^j \\
&= \sum_{j=1}^{n} A_j \left(2\sqrt{p(1-p)}\right)^j \\
&= \sum_{j=0}^{n} A_j \left(2\sqrt{p(1-p)}\right)^j - A_0 \\
&= A\left(2\sqrt{p(1-p)}\right) - 1
\end{aligned}$$

where line $*$ above follows from the fact that $(1-p)^{j-i}p^i$ is a decreasing function of i if $p < \frac{1}{2}$ (as it is for any reasonable channel). Then $*\,*$ follows too because $\lceil \frac{j}{2} \rceil \ge \frac{j}{2}$. [Ex 37–39]

6.9 Exercises for Chapter 6

1. Use the method of proving Theorem 6.1 to find a member of Ham(2, 5).
2. Show that the suggested column selection method does give a p.c. matrix of a member of Ham(r, q).
3. Show that any member of Ham(r, q) must have a p.c. matrix containing columns $\boldsymbol{x}$, $\boldsymbol{y}$, $\boldsymbol{z}$ whose first two members are $a0, 0b, cc$ respectively, and the rest zero, for some non-zero a, b, c.
4. For $C \in$ Ham(5, 2) with the suggested column ordering decode the received word $\boldsymbol{r}$ with 1 in the first four positions and zeros elsewhere.
5. Find the 'convenient decoding' form for a p.c. matrix of an [8, 6, 3] Hamming code C, and use it to decode 12312300.
6. Show that

$$G = \begin{bmatrix} 1 & 1 & 1 & 1 & 1 & 1 & 1 \\ 1 & 0 & 0 & 0 & 1 & 0 & 1 \\ 1 & 1 & 0 & 0 & 0 & 1 & 0 \\ 0 & 1 & 1 & 0 & 0 & 0 & 1 \end{bmatrix}$$

generates a binary Hamming code.

7. Calculate the error-correcting capability of the dual of any code C in Ham (r, q).
8. Show that no linear equidistant ternary code C which is 2-error correcting and has dimension 4, length 10 can exist.
   [Hint: Imagine the code to be written out as an 81×10 array and, using Theorem 5.4 count the total number of non-zeros in the whole code in two ways.]
9. Show that all binary simplex codes are optimal.
10. Find upper and lower bounds for the size of the best binary linear 2-error correcting code of length 12.
11. In the construction used in our proof of Theorem 6.9, show that k is the dimension of the code constructed.
12. Show that any binary code with $d = n$ is equivalent to a repetition code.
13. Show that perfect binary repetition codes must have odd length and that there are no perfect non-binary repetition codes with $n > 1$.
14. Our definition of $Res(C, \mathbf{c})$ seems to depend on G as well as on $\mathbf{c}$. Show that this is not the case.
15. Let $w(\mathbf{c}) = w(\mathbf{c}')$ for distinct codewords $\mathbf{c}, \mathbf{c}'$ of C. Show that in general $Res(C, \mathbf{c}) \neq Res(C, \mathbf{c}')$, and indeed these two residual codes need not be equivalent, nor even have the same dimension.
16. Why must a $\mathbf{c}$ with the properties specified in Definition 6.3 exist?
17. For each real number x prove that
$$\left\lceil \frac{\lceil x \rceil}{2} \right\rceil = \left\lceil \frac{x}{2} \right\rceil .$$
18. With generator matrices chosen as in the preamble to Theorem 6.11 show that
$$d(Res^i C) \geq \left\lceil \frac{d}{2^i} \right\rceil$$
   where $Res^2\, C$ means $Res(ResC) \cdots$ etc.
19. Do a similar analysis comparing the performances of the Griesmer, Hamming and Plotkin bounds in upper bounding the size of $[20, k, 11]$ binary codes.
20. Show that all binary simplex codes with dimension ≥ 3 satisfy the conditions of Theorem 6.14 and are therefore self-orthogonal.
21. Explain why the previous two theorems do not generalize to the fields Z_p with $p > 3$.
22. Give an example to show that Theorem 6.18 does not extend to Z_p with $p > 3$.
23. Find a self-dual Hamming code.

24. Find a binary [10, 5] self-dual code.
25. Show that if a binary self-orthogonal code C has a generator matrix G in which every row has a weight which is a multiple of 4 then every codeword weight is a multiple of 4. [Hint : use the formula, $w(\boldsymbol{x}+\boldsymbol{y}) = \cdots$.]
26. Check that H does indeed produce the codewords listed, and that the listings of C and D have the properties claimed.
27. Show that E' has 2^{12} codewords.
28. Derive a formula for $w(\boldsymbol{x}+\boldsymbol{y}+\boldsymbol{z})$ from the formula for $w(\boldsymbol{x}+\boldsymbol{y})$.
29. Show that for each $\boldsymbol{a}, \boldsymbol{b}$ in C', $\boldsymbol{a} \odot \boldsymbol{b}$ has even weight.
30. Show that the Hamming code with

$$H = \begin{bmatrix} 1 & 0 & 0 & 1 & 1 & 1 & 0 \\ 0 & 1 & 0 & 1 & 1 & 0 & 1 \\ 0 & 0 & 1 & 1 & 0 & 1 & 1 \end{bmatrix}$$

is not cyclic.
31. Prove the second equality in our statement of the MacWilliams identity.
32. Show that there is no-error correcting code which is the dual of a repetition code.
33. Use the MacWilliams identity to prove that the dual of the binary repetition code of length n consists of all the even weight words of Z_2^n. Establish the result by an alternative method.
34. Verify from this expression that $A_1 = A_2 = 0$, so that Ham (r, q) is 1-error correcting.
35. Explain how these expressions are obtained.
36. Use this recurrence relation to evaluate A_3, A_4 and A_5 for Ham(4, 2).
37. Show that for Ham(3, 2) with $p = 0.01$ the actual probability of a decoding error is much smaller than the upper bound just derived.
38. Let C be any binary code and C' its extension by adding on overall parity check. How are their weight distributions A_i and A_i' related?
39. A linear code C over Z_5 has generator matrix

$$G = \begin{bmatrix} \boldsymbol{r}_1 \\ \boldsymbol{r}_2 \end{bmatrix} = \begin{bmatrix} 1 & 0 & 4 & 2 & 3 & 1 \\ 0 & 1 & 4 & 1 & 0 & 2 \end{bmatrix}.$$

By considering the weights of the codewords $\boldsymbol{r}_2$ and $\boldsymbol{r}_1 + \lambda \boldsymbol{r}_2, \lambda = 0, 1, 2, 3, 4$, find the weight distribution of C.
[Hint : consider multiples of the codewords above, and the fact that C has 25 codewords.] Apply the MacWilliams identity to obtain the weight distribution of $C^{\perp}$.

7

Polynomials for codes

Chapter 3 introduced some basic number theory so that we had some useful mathematical machinery to deal with linear codes, principally through the properties of the fields Z_p. This chapter has a similar motivation, and will be put to use in the next chapter on cyclic codes.

7.1 The first definitions

A *polynomial* in a single variable is an expression of the form $a_0 + a_1x + a_2x^2 + \cdots + a_nx^n$ where n is a positive or zero integer and the coefficients $a_0, a_1, \cdots, a_n$ will be restricted to being elements of a field.

$F[X]$ denotes the set of all polynomials in a single variable with coefficients in the field F.

For each $f \in F[X]$ the largest n for which x^n has a non-zero coefficient is called the *degree* of f, denoted by $\deg(f)$.

If $\deg(f) = n$ and $a_n = 1$ then f is called a *monic* polynomial.

If $\deg(f) = n$ and $f(x) = a_nx^n + \cdots$, then a_nx^n is called the *leading term* of f.

The *zero polynomial* is the one in which all coefficients are zero. In this case the definition of degree given above does not work so $\deg(f)$ is then defined to be 'minus infinity'. This is a purely conventional (but useful) definition. We shall use 0 to denote both the zero of F and the zero polynomial, and rely on context or emphasis to avoid confusion.

Polynomials with degrees 0 and 1 are called *constant* and *linear* polynomials respectively.

We sometimes use f and sometimes $f(x)$ for a member of $F[X]$. Often it is only the sequence of coefficients of f which is of any significance, so the f notation is appropriate here. On other occasions we shall 'evaluate the polynomial' by substituting a member of F for x in $a_0 + a_1x + \cdots + a_nx^n$, and for this $f(x)$ is appropriate. We resist the temptation to make the distinction between the two notations rigorous, and let convenience decide between them.

The polynomials $a_0 + a_1x + \cdots + a_nx^n$ and $b_0 + b_1x + \cdots + b_nx^n$ are called *equal* if they have the same degree and $a_i = b_i$ for all i. The usual $=$ sign will denote both equality of polynomials f, g, and equality of field elements $f(\alpha)$ and $f(\beta)$.

7.2 Operations in $F[X]$

We assume you are familiar with adding, subtracting and multiplying members of $F[X]$, and note that $F[X]$ is closed under these operations. Division is a different matter: if f and g are in $F[X]$ then in general $f \div g$ is not in $F[X]$. But this is not an unfamiliar problem : $8 \div 3$ is not an integer, but by using the idea of a remainder it is possible to discuss division of integers entirely in terms of integers. The result of dividing 8 by 3 is $8 = 2 \times 3 + 2$. Our main purpose now is to demonstrate an analogue of the division algorithm of Chapter 3 for polynomials. In the integer version there was a uniqueness clause : in dividing n by $m(m \neq 0)$ and expressing the result as $n = qm + r$, restricting r to the range $0 \leq r < m$ makes q and r unique. There is no such natural notion of 'size' of polynomials to help us, but deg (f) *is* an integer and we can use this to restrict the remainder polynomial. [Ex 1]

Theorem 7.1 Let f, g be members of $F[X]$ with $g \neq 0$. Then there are polynomials q, r in $f[X]$ such that $f = qg + r$, and q, r are unique if r is restricted by the condition $\deg(r) < \deg(g)$.

Proof. Our proof is not the most elegant but it has the virtue of being constructive – that is, one which contains a method of finding q and r.

Let $\deg(f) = n$, $\deg(g) = m$, $f(x) = a_nx^n + \cdots + a_0, g(x) = b_mx^m + \cdots + b_0$.

We first dispose of the case $\deg(f) < \deg(g)$. We require $f = qg + r$, and suppose $q \neq 0$. Then $\deg(q\ g) \geq \deg(g)$ and $\deg(r) < \deg(g)$, so $\deg(qg+r) \geq \deg(g)$, but this contradicts $\deg(f) < \deg(g)$. Hence $q = 0$ in this case and $f = r$. [Note how this argument has used Exercise 1.]

For the case $\deg(f) \geq \deg\ (g)$ q cannot be the zero polynomial, and its degree is $n-m$. So let $q(x) = q_{n-m}x^{n-m} + \cdots + q_0$ and $r(x) = r_{m-1}x^{m-1} + \cdots + r_0$ (where some of the r_i, including r_{n-m-1}, could be zero). Writing out $f = qg + r$ we have

$$a_nx^n + \cdots + a_0 = (q_{n-m}x^{n-m} + \cdots + q_0)(b_mx^m + \cdots + b_0) \\ +(r_{m-1}x^{m-1} + \cdots + r_0)$$

Equating the coefficients of $x^n, x^{n-1}, \cdots, x^m$ gives the following set of equations in F:

$$\begin{array}{lclclcl}
a_n & = & b_m q_{n-m} & & & & \\
a_{n-1} & = & b_m q_{n-m-1} & + & b_{m-1} q_{n-m} & & \\
a_{n-2} & = & b_m q_{n-m-2} & + & b_{m-1} q_{n-m-1} & + & b_{m-2} q_{n-m} \\
 & \vdots & & & & & \\
a_m & = & b_m q_0 & + & b_{m-1} q_1 & + & b_{m-2} q_2 + \cdots + b_{2m-n} q_{n-m}
\end{array}$$

Notice that not all these terms mentioned above need exist. For example, in the last equation, if $n > 2m$, then the last term and possibly some earlier ones will not appear. If we interpret any of these non-existent terms as zeros we can retain the equations in the form given. What *is* important is that $b_m \neq 0$, so the first equation gives the unique solution $q_{n-m} = b_m^{-1} a_n$. This can be substituted into the second to get $q_{n-m-1} = b_m^{-1}(a_{n-1} - b_{m-1} q_{n-m})$, then the third gives q_{n-m-2}, and so on. Hence all the qs are uniquely determined from these equations.

Each of the remaining equations, obtained by equating the coefficients of $x^{m-1}, x^{m-2}, \cdots, x^0$ contain only $r_{m-1}, r_{m-2}, \cdots, r_0$ respectively, with some of the qs (which have been found previously). Hence these equations will determine all the rs uniquely. □

[Ex 2]

There is a method of writing down the calculation in Exercise 2 in a form which is reminiscent of 'long division' of integers. We illustrate this with the first example from Exercise 2.

$$\begin{array}{cl}
 & 3x^6 + 2x^5 + 0x^4 + 4x^3 + 0x^2 + 2x + 2 \\
= & \underline{2x^2}(4x^4 + x^3 + x^2 + 3x + 1) + r_1(x) \\
= & 3x^6 + 2x^5 + 2x^4 + x^3 + 2x^2 + r_1(x)
\end{array}$$

So

$$\begin{array}{rcr}
r_1(x) & = & 3x^6 + 2x^5 + 0x^4 + 4x^3 + 0x^2 + 2x + 2 \\
 & & \underline{-(3x^6 + 2x^5 + 2x^4 + x^3 + 2x^2)} \\
 & = & -2x^4 + 3x^3 - 2x^2 + 2x + 2
\end{array}$$

- -A

So

$$\begin{array}{rcl}
r_1(x) & = & \underline{2}(4x^4 + x^3 + x^2 + 3x + 1) + r_2(x) \\
 & = & -2x^4 + 2x^3 + 2x^2 + x + 2 + r_2(x)
\end{array}$$

So

$$\begin{aligned} r_2(x) &= -2x^4 + 3x^3 - 2x^2 - 2x + 2 \\ &\quad \underline{-(-2x^4 + 2x^3 + 2x^2 + x + 2)} \\ &= x^3 + x^2 + x \end{aligned}$$

- -B

$$\begin{aligned} &3x^6 + 2x^5 + 0x^4 + 4x^3 + 0x^2 + 2x + 2 \\ &\quad = (2x^2 + 2)(4x^4 + x^3 + x^2 + 3x + 1) + (x^3 + x^2 + x). \end{aligned}$$

Don't forget that all the numerical calculation is carried out modulo 5.

The same calculation in a shorter 'long division' format is shown below.

```
                                           :A       :B
                                      2x^2 :    + 2 :     quotient
                        ______________________________
4x^4 + x^3 + x^2 + 3x + 1 ) 3x^6 + 2x^5 + 0x^4 + 4x^3 + 0x^2 + 2x + 2
                            3x^6 + 2x^5 + 2x^4 + x^3 + 2x^2
                          ______________________________________
                                  - 2x^4 + 3x^3 - 2x^2 + 2x + 2
                            - - - - - - - - - - - - - - - - - - - - - A
                                  - 2x^4 + 2x^3 + 2x^2 + x + 2
                                 ________________________________
                                        x^3 + x^2 + x              remainder.
```

[Ex 3]

For polynomial division in $Z_p[X]$ with p greater than about 5 a useful preliminary step is to draw up a table of inverses in Z_p. For example, in doing the first step in $Z_{11}[X]$ of the calculation

$$(5x^{11} + \cdots) = (ax^4 + \cdots)(8x^7 + \cdots) + r(x)$$

we have to find a to satisfy $8a \equiv 5 \bmod 11$, so $a = 5 \times 8^{-1} = 5 \times 7 = 2$.

7.3 Factorization in $Z_p[X]$

If f, g, h are polynomials in $F[X]$ and $f = gh$ then g and h are called divisors or factors of f. Equivalently, f is a multiple of g (and of h). These statements are equivalent to the remainder polynomial on dividing f by g being zero. We continue to use the notation $g|f$ from Chapter 3.

$F[X]$, just like Z, has a unique prime factorization theorem, but before stating it we need to define the analogues of primes in $F[X]$.

Definition 7.1 $f \in F[X]$ is called *irreducible* if deg $(f) > 0$ and f is not the product of two polynomials both having positive degree.

The essence of this definition is that f is irreducible if it has no non-trivial factorization. Factorization in $F[X]$ is in general a hard problem with applications to coding theory and many other areas. Linear polynomials are clearly irreducible, and there is an easy test for reducibility of quadratic and cubic polynomials over finite fields. The test is a consequence of the following result.

Theorem 7.2 Let $f(x)$ and the linear polynomial $x - \alpha$ be members of $F[X]$. Then $x - \alpha | f(x)$ if and only if $f(\alpha) = 0$.

Proof.

(i) $x - \alpha | f(x) \Rightarrow f(x) = q(x) \cdot (x - \alpha)$ for some $q \in F[X]$
$\Rightarrow f(\alpha) = 0$

(ii) Divide $f(x)$ by $x - \alpha$ to get $f(x) = q(x) \cdot (x - \alpha) + r(x)$ where, by Theorem 7.1, $r(x)$ is either zero or has degree 0.

$$\text{Then}\quad f(\alpha) = 0 \Rightarrow q(\alpha) \cdot 0 + r(\alpha) = 0 \Rightarrow r = 0 \Rightarrow x - \alpha | f(x) \qquad \square$$

[Ex 4]

Now by Exercise 1 again, if a quadratic or cubic polynomial has a non-trivial factorization it can only be into a pair of linear factors, or a linear and a quadratic (or three linears) respectively. Furthermore, by Exercise 5 below, we need only check whether any of the *monic* linear polynomials are factors. [Ex 5,6]

At the price of working a little harder quartic and quintic polynomials over smallish fields can also be investigated, since if such a polynomial has no linear factors then its only possible factorisations are into a pair of quadratics or a quadratic and cubic respectively. [Ex 7]

Theorem 7.3 Each non-constant member of $F[X]$ is either irreducible or is the product of a constant with a unique family of irreducible monic polynomials. $\square$

We shall not prove this, but simply point out that its proof and much of the accompanying theory is identical to the corresponding theory for integers in Chapter 3. Other similarities between Z and $F[X]$ are outlined in the remainder of this section.

For f, g not both zero in $F[X]$ we can define a greatest common divisor of f and g as a polynomial of maximal degree which is a divisor of both. The *gcd* is unique up to constant multiples (see Exercise 5). $gcd(f,\ g)$ is any polynomial of minimal degee (but ≥ 0) in the set $\{sf + tg : s \in F[X],\ t \in F[X]\}$, and $gcd(f, g)$ may be found by a process analogous to Euclid's algorithm. Any pair of polynomials whose *gcd* is a constant is called a relatively prime pair.

A useful analogue of Euclid's lemma also holds : if i is irreducible and $i|fg$, then $i|f$ or $i|g$.

An example of the $F[X]$ version of Euclid's algorithm and further practice in polynomial division is given in the exercise below. [Ex 8]

7.4 Congruence of polynomials

If f, g, h are all polynomials in $F[X]$ then we say that f is congruent to g modulo h (written $f \equiv g \mod h$) if $f - g$ is a multiple of h. So the idea and notation is exactly the same as for integers. For our work on cyclic codes the significant result of a polynomial division will be the remainder rather than the quotient, so congruence will clearly be important. All the familiar properties of integer congruence hold also for polynomial congruence, which justifies the technique illustrated by the following example.

Suppose in $Z_7[X]$ we require the polynomial of smallest degree which is congruent to $3x^4 + 5x^3 + 2x^2 + 4 \mod 2x^2 + 3x + 1$. In other words, what is the remainder when the first of these is divided by the latter? From Theorem 7.1 we know the answer has degree at most 1, and we can find it *without* carrying out the division as follows. Working modulo $2x^2 + 3x + 1$ throughout we have

$$2x^2 \equiv -3x - 1$$

so

$$x^2 \equiv 2^{-1}(-3x - 1) = 4(4x + 6) = 2x + 3$$

Hence

$$x^3 = x(x^2) \equiv x(2x + 3) = 2x^2 + 3x \equiv 2(2x + 3) + 3x = 7x + 6 \equiv 6$$

and

$$x^4 = x(x^3) \equiv x(6) = 6x$$

so

$$3x^4 + 5x^3 + 2x^2 + 4 \equiv 3(6x) + 5(6) + 2(2x + 3) + 4 \equiv x + 5$$

[Ex 9]

7.5 Rings and ideals

The theory of cyclic codes depends on properties of certain subsets of $F[X]$, but first we discuss the main ideas in a more familiar context – the integers. Algebraically the most important thing about Z is how it behaves with respect to the operations of addition and multiplication. The structure $(Z, +, \times)$ is an example of what is called a *ring*. Any set R, on which two binary operations (called for convenience $+$ and $\times$) are defined, is a ring if it satisfies the following conditions for arbitrary choices of r, s, t in R:

1. $r + s \in R \qquad r \times s \in R,$
2. $r + s = s + r$

3. $(r+s)+t=r+(s+t)$, $(r \times s) \times t = r \times (s \times t)$;
4. $r \times (s+t) = (r \times s) + (r \times t)$, $(s+t) \times r = (s \times r) + (t \times r)$;
5. R has a member, usually called 0, such that $r + 0 = r$;
6. R has a member, usually called $-r$, such that $r + (-r) = 0$.

If you have met some abstract algebra you will recognise that this amounts to saying that R is a commutative (or Abelian) group under $+$, $\times$ is associative, and $\times$ is left and right distributive over $+$.

If $\times$ is also commutative ($r \times s = s \times r$ for all r, s in R), and R has a member (usually called 1 or unity) such that $1 \times r = r$ for each r in R, then $(R, +, \times)$ is called a *commutative ring with unity.* $(Z, +, \times)$ is clearly a ring of this type, as is $(F[X], +, \times)$ where in this case the operations are polynomial addition and multiplication. [Ex 10]

Now let us return to Z, our first example of a ring, and think about the subset of all even integers which we denote by $2Z$. This is also a commutative ring (check the defining properties), but this time without a unity. Any subset of a ring R, which is also a ring in its own right, is called a *subring* of R. But our example has the additional property that if e, x are any members of $2Z$, Z respectively then ex is also in $2Z$. This leads to the following definition:

Definition 7.2 Let $(R, +, \times)$ be a ring with subring $(S, +, \times)$. S is called a (two-sided) ideal of R if for all r in R and all s in S, S contains $r \times s$ and $s \times r$.

If R is not commutative then it could happen that $r \times s \in S$ but $s \times r \notin S$. In this case we would have to distinguish between left, right and two-sided ideals. However, from now on all our rings will be commutative so we may drop the '(two-sided)' and 'and $s \times r$' from Definition 7.2.

Clearly in our example the ideal $2Z$ is simply the set of all integer multiples of a particular member 2 (or of -2), and Z is special in this respect : all its ideals are of this type.

Theorem 7.4 If I is an ideal of Z then there is some particular integer m such that $I = \{mx : x \in Z\}$.

Proof. One possibility is that $I = \{0\}$ (check that this really is an ideal), and in this case $m = 0$ satisfies the claim of the theorem. So now suppose I has at least one non-zero member i. I is a subring of Z so $i^2 \in I$, and of course $i^2 > 0$. Let m be the smallest positive member of I. By the ring properties I must contain every multiple of m, so we just have to show that it contains nothing else. To do this, let j be any member of I, and divide j by m to get $j = mq + r$ with $0 \leq r < m$. Using the properties of ideals we have

$$(j \in I \text{ and } m \in I) \Rightarrow mq \in I \Rightarrow -mq \in I \Rightarrow j - mq \in I \Rightarrow r \in I.$$

Hence r must be 0 so $j = mq$ □

For the remainder of this book all our rings will be commutative with unity. With this understanding we make the following definition.

Definition 7.3 Any ideal I of a ring R which consists of all the multiples of a specific member of R by every member of R is called a *principal ideal* of R.

We have just shown that all ideals of Z are principal. The specific member referred to above is called a *generator* of I, so $2Z$ has just 2 and -2 as its generators. Note that Z itself is also an ideal of Z, and in general any ring is an ideal of itself.

The ideas and proof of Theorem 7.4 apply when Z is replaced by $F[X]$. You should carry out the proof in this case : the only change you will have to make is that m becomes any non-zero polynomial of smallest degree in I, and then r will be the remainder polynomial, with degree smaller than that of m. [Ex 11]

Finally, (and not to be used later), you may have wondered whether some rings have non-principal ideals, and whether examples are easy to construct.

If so, try. [Ex 12]

7.6 Exercises for Chapter 7

1. $f, g \in F[X]$. What is the relation between $\deg(f)$, $\deg(g)$, $\deg(f+g)$, $\deg(f-g)$ and $\deg(fg)$?
2. Find q and r when $3x^6 + 2x^5 + 4x^3 + 2x + 2$ in $Z_5[X]$ is divided by
   (a) $4x^4 + x^3 + x^2 + 3x + 1$,
   (b) $2x^6 + x^4 + 3x^2$,
   (c) $3x + 4$.
3. Use long division to find q and r if
   (a) $x^6 + x^3 + x^2 + x = q(x)(x^4 + x^2 + x + 1) + r(x)$ in $Z_2[X]$,
   (b) $x^6 + 2x^5 + 2x + 1 = q(x)(2x^3 + x^2 + x + 1) + r(x)$ in $Z_3[X]$.
4. Show that $1 + x | g(x)$ in $Z_2[X]$ if and only if $g(x)$ has an even number of non-zero terms.
5. Show that, for $f \in Z_p[X]$, for each $\beta \in Z_p \backslash \{0\}$, $x - \alpha | f(x)$ if and only if $\beta(x - \alpha) | f(x)$.
6. Use the result of Exercise 5 to factorize each of the following into a constant and monic irreducibles.
   (a) $3x^2 + 4x + 3$ in $Z_5[X]$,
   (b) $2x^3 + 2x^2 + x + 2$ in $Z_3[X]$,

(c) $x^3 + 4x^2 + x + 1$ in $Z_7[X]$,
(d) $x^3 + 2x^2 + 2$ in $Z_3[X]$,
(e) $2x^2 + 3x + 4$ in $Z_7[X]$.

7. Determine whether the quintic polynomial $f(x) = (x+1)(x+2)^2(x+3)^2 + 4$ in $Z_5[X]$ is irreducible, and if not factorize it completely.

8. Apply a polynomial version of Euclid's algorithm to find $gcd(f, g)$ where $f(x) = x^{12} + x^4 + x^3 + x^2 + x + 1$, $g(x) = x^8 + 2x^6 + x^5 + x^2 + 2x + 2$, both in $Z_3[X]$.

9. Find the quadratic polynomial congruent to $x^7 + 2x^6 + 2x^4 + x^3 + x^2 + 2$ mod $x^3 + 2x$

   (a) in $Z_3[X]$,
   (b) in $Z_5[X]$.

10. For each of the following cases determine whether $(A, +, \times)$ is

    (a) a ring,
    (b) a ring with unity,
    (c) a commutative ring.

    In (i), (ii) and (iii) + and $\times$ are ordinary matrix addition and multiplication.

    (i) $A =$ the set of all 3×3 matrices with entries in Z.
    (ii) $A =$ the set of all 2×2 matrices with entries in Z, of the form

$$\begin{bmatrix} a & 0 \\ b & c \end{bmatrix}.$$

    (iii) As for (ii), but of the form

$$\begin{bmatrix} 0 & x \\ 0 & y \end{bmatrix}.$$

    (iv) $A =$ the set of all finite subsets of N, $+$ is $\oplus$, the symmetric difference or 'exclusive or' operator, and $\times$ is $\cap$.

11. Let $m(x)$ be a generator of the principal ideal I of $F[X]$. Show that for each $\alpha \in F\backslash\{0\}$, $\alpha m(x)$ is also a generator of I. Show further that if $m(x)$ is a
generator of minimal degree, then there are no generators of this degree except the constant non-zero multiples of $m(x)$.

12. Show that $F[X, Y]$, the set of polynomials in two variables with coefficients in a field F, is a ring, and that the subset $I = \{xs(x, y) + yt(x, y) : s \in F[X, Y], t \in F[X, Y]\}$ is an ideal, but not a principal ideal.

8

Cyclic codes

8.1 Introduction

The aim of this chapter is to show the connection between cyclic codes introduced in section 6.7 and the polynomial algebra of the previous chapter. Then we exploit the latter to throw light on the former. The connection is that each word in F^n, say $a_0a_1\cdots a_{n-1}$, is thought of as a polynomial $a_0+a_1x+\cdots+a_{n-1}x^{n-1}$ in $F[X]$. Notice that in order to retain the sensible notation of a_i being the coefficient of x^i we have to represent the words as $a_0a_1\cdots a_{n-1}$ rather than $a_1a_2\cdots a_n$. Of course $F[X]$ has infinitely many members, whereas F^n (e.g. Z_p^n) has only p^n members, so the match is not perfect. To overcome this we just replace the set $F[X]$ by its subset of polynomials with degree less than n. Then it is easy to see that if $\mathbf{c}_1$ and $\mathbf{c}_2$ are any codewords and $c_1(x), c_2(x)$ are their corresponding polynomials and λ_1, λ_2 are any members of F, then the codeword $\lambda_1\mathbf{c}_1 + \lambda_2\mathbf{c}_2$ has $\lambda_1 c_1(x)+\lambda_2 c_2(x)$ as its polynomial representative. For cyclic codes it turns out that multiplication of polynomials is a very useful operation, but here we hit a snag: if C has length n so that its polynomials have degree less than n, the product of two such polynomials can have degree at least n, so does not represent a codeword. To get round this we pick a polynomial f of degree n and define our multiplication operation to be polynomial multiplication modulo $f(x)$, so that any two polynomials which differ by a multiple of $f(x)$ are regarded as equal. So our set of polynomials of degree less than n which will represents codewords can be thought of as the set of all remainder polynomials on division by $f(x)$. Our notation for this set is $F[X]/f(x)$, and we shall make use of congruence notation by using $\equiv$ for equality in $f[X]/f(x)$ and $=$ for equality in $F[X]$. [Ex 1, 2]

8.2 The choice of modulus

All our remarks about representing codes as sets of polynomials have been very general. So what of cyclic codes? Let C be a cyclic $[n, k]$ code over F

and $\mathbf{c} = a_0 a_1 \cdots a_{n-1}$ one of its codewords. $\mathbf{c}^1$, the first cyclic shift of $\mathbf{c}$, is also a codeword, so let us compare the polynomials which represent these words:

$$c(x) = a_0 + a_1 x + a_2 x^2 + \cdots + a_{n-2}x^{n-2} + a_{n-1}x^{n-1}$$

and

$$c'(x) = a_{n-1} + a_0 x + a_1 x^2 + \cdots + a_{n-3}x^{n-2} + a_{n-2}x^{n-1}.$$

Observe that $c'(x)$ can be obtained (nearly) by multiplying $c(x)$ by x. In fact the only difference is that $c'(x)$ has the term a_{n-1} whereas $xc(x)$ has the term $a_{n-1}x^n$. So $c'(x) \equiv xc(x)$ provided we choose f so that $x^n \equiv 1 \bmod f(x)$. In other words, let $f(x)$ be $x^n - 1$, and then we have the very neat result that cyclic shift in C corresponds to multiplying by x in $f[X]/x^n - 1$. For this reason we make this choice of f for the polynomial representation of any cyclic code of length n over F.

There is also something very special about the precise set of polynomials which represent the codewords:

Theorem 8.1 A set I of polynomials in the ring $R = f[X]/x^n - 1$ represents a cyclic code C if and only if I is an ideal of R.

Proof. Let I be an ideal and let c_1, c_2 be members of I, and λ any constant member of R. Then $c_1 + c_2 \in I$ by subring property 1 of I, and $\lambda c_1 \in I$ by the ideal property. In terms of the code C, we have proved C is closed under addition and scalar multiplication. In other words C is linear. To show that it is cyclic, $x \in R$ so $xc(x) \in I$ by the ideal property again. As we have seen this means that C is closed under the cyclic shift operation. Hence C is a cyclic code.

Conversely, let C be cyclic. We have to show that I has all the properties 1–6 of section 7.5 and has the ideal property. Properties 2, 3, 4 hold for all members of R so in particular they apply to I (we say that I inherits these properties from R). C is linear so $\mathbf{0} \in C$ and $\mathbf{c} \in C \Rightarrow (-1)\mathbf{c} \in C$, which translates into properties 5 and 6 for I. The addition part of property 1 also follows from the linearity of C. Now for the ideal property : let $\mathbf{c}$ be any codeword, represented by $c(x)$ in I, and let $p(x) = p_0 + p_1 x + \cdots + p_{n-1}x^{n-1}$ be any polynomial in R. Then $p(x)c(x) = p_0 c(x) + p_1 xc(x) + \cdots + p_{n-1}x^{n-1}c(x)$, and this represents the word $p_0\mathbf{c} + p_1\mathbf{c}^1 + p_2\mathbf{c}^2 + \cdots + p_{n-1}\mathbf{c}^{n-1}$, which is a linear combination of cyclic shifts of $\mathbf{c}$, so must be in C. Hence $p(x)c(x) \in I$ so I has the ideal property.

The only remaining item is the multiplicative part of property 1. But we have already done the work for this: $p(x)c(x) \in I$ for all $p \in R, c \in I$, so in particular $p(x)c(x) \in I$ for all $p \in I, c \in I$. □

From Exercise 2 the previous theorem tells us that C is a cyclic code if and only if its set of representative polynomials in R is the set of all multiples of some single polynomial, and conversely, every such set of polynomials represents a cyclic code. We use the notation $I = \langle p \rangle$ for the ideal consisting of all multiples of p, and p is called a generator of I. Notice that we say '*a*' generator because we can have $\langle p \rangle = \langle q \rangle$, for distinct p, q, so the ideals of R need not be as numerous as you may have thought. Indeed it is clear that $\langle p \rangle = \langle \lambda \rho \rangle$ for any non-zero constant λ, and the next exercise shows that p and q can differ by more than just a constant factor, yet still generate the same ideal. [Ex 3]

Note also that Exercise 3 emphasizes something you may have observed already – that the property deg(fg) =deg(f)+ deg(g) which holds in $F[X]$ fails in $F[X]/f(x)$.

8.3 Generator matrices and generator polynomials

We have seen that linear codes have generator matrices and the subclass of cyclic codes have generator polynomials. The aim of this section is to connect these two ideas of 'generator'. But first a theorem:

Theorem 8.2 Every cyclic $[n, k]$ code C, other than $\{\mathbf{0}\}$, has a generator matrix G of the form

$$\begin{bmatrix} g_0 & g_1 & g_2 & \cdots & g_t & 0 & 0 & & \cdots & 0 \\ 0 & g_0 & g_1 & \cdots & g_{t-1} & g_t & 0 & & \cdots & 0 \\ & & & & \vdots & & & & & \\ & & & & \vdots & & & & & \\ 0 & 0 & \cdots & 0 & g_0 & g_1 & \cdots & \cdots & \cdots & g_r \end{bmatrix}$$

in which the last $k-1$ places of row 1 are zeros, each row is the first cyclic shift of the previous row, and neither g_0 nor g_t is zero.

Proof. First observe that this matrix has the right number of rows, which is k, the dimension of C. If the first row is a codeword of C, then so are all the rest by the cyclic property. It remains to show, then, that the rows are independent and that C has a codeword of the form given by row 1. (Establishing the first of these assertions will be your contribution to the proof.) [Ex 4]

Let $\{\mathbf{c}_1, \mathbf{c}_2, \cdots, \mathbf{c}_k\}$be any basis for C. We claim that there is some choice of constants $\lambda_1, \lambda_2, \cdots, \lambda_k$ (other than all zero) for which $\lambda_1\mathbf{c}_1 + \lambda_2\mathbf{c}_2 + \cdots + \lambda_k\mathbf{c}_k$ is a word ending in $k-1$ zeros. Let $\mathbf{c}'_i$ be the word consisting of just the last $k-1$ digits of $\mathbf{c}_i$. Then $\mathbf{c}'_1, \mathbf{c}'_2, \cdots, \mathbf{c}'_k$ are k words in F^{k-1}, a space of dimension $k-1$. Hence these words are dependent so there are $\lambda_1, \lambda_2, \cdots \lambda_k$, not all zero, such that

$$\lambda_1\mathbf{c}'_1 + \lambda_2\mathbf{c}'_2 + \cdots + \lambda_k\mathbf{c}'_k = \mathbf{0}',$$

So

$$\lambda_1\mathbf{c}_1 + \lambda_2\mathbf{c}_2 + \cdots + \lambda_k\mathbf{c}_k$$

is a word of the form claimed.

Finally, if $g_0 = 0$, then the first column of G is all zero, so every codeword of C has zero as its first place. But this is impossible because, by hypothesis, C has non-zero words so some cyclic shift of such a word will have a non-zero digit as its first symbol. Similarly, by considering the last digit of the codewords, g_r cannot be zero. □

The g_i in the theorem above will turn out to be the coefficients of a generator polynomial for C (or strictly, the ideal of $F[X]/x^n - 1$ corresponding to C, though we shall wilfully ignore the distinction). But which generator? You have seen in Exercise 3 that they are not unique, so we first prove a uniqueness result.

Theorem 8.3

(a) Of all non-zero members of an ideal I those of smallest degree are simply constant multiples of each other.

(b) Each of the minimal degree members of I is a generator of I.

Proof.

(a) Suppose not. That is, let $a(x) = a_l x^l + \cdots$ and $b(x) = b_l x^l + \cdots$ be two such members with b not a constant multiple of a. Then $a(x) - a_l b_l^{-1} b(x)$ is non-zero, with degree $< l$. But $-a_l b_l^{-1} \in R$, so by the ideal properties $a(x) - a_l b_l^{-1} b(x) \in I$, which contradicts the minimality hypothesis.

(b) For all $r \in R, ra \in I$, so $\langle a \rangle \subseteq I$. To prove the reverse inclusion, let i be any member of I. Divide $i(x)$ by $a(x)$ to get $i(x) = a(x)q(x) + r(x)$ with $\deg(r) < \deg(a)$. Note that $=$ rather than the weaker $\equiv$ is correct here because $\deg(i) < n$.

By a similar argument to that used in (a) above, $r(x) \in I$ so $r(x) = 0$. Hence $a(x)|i(x)$ so $I \subseteq \langle a \rangle$ and (b) follows. □

Corollary. There is only one monic polynomial of minimal degree in I. □

As a result of the last theorem and its corollary we can make the following definition.

Definition 8.1 The unique monic polynomial of smallest degree in an ideal I is called the generator of I.

Now we can relate our two theorems of this section.

Theorem 8.4 A generator matrix for the cyclic $[n, n-t]$ code C of the form established in Theorem 8.2 is obtained by taking $g_0, g_1, \cdots, g_t$ to be the coefficients in the generator polynomial of C.

Proof. Let the G of Theorem 8.3 generate C. Then $g(x) = g_0 + g_1 x + \cdots + g_r x^t \in C$,so $\langle g \rangle \subseteq C$. But every codeword is a linear combination of $g(x), xg(x), \cdots x^{n-t-1} g(x)$.

That is $c(x) = \lambda_0 g(x) + \lambda_1 x g(x) + \cdots + \lambda_{n-t-1}x^{n-t-1}g(x) \in \langle g \rangle$.
Therefore $C \subseteq \langle g \rangle$, so $C = \langle g \rangle$.

To show that g can be taken as the generator of C, suppose there is a polynomial g' of smaller degree, say $t - \alpha$, in $I(\alpha > 0)$. Then we have the contradiction that the $[n, n-t]$ code C has $n-t+\alpha$ independent codewords, $g', xg', x^2g', \cdots, x^{n-t+\alpha-1}g'$.

So g is a generator of minimal degree, and if we take the monic constant multiple of g the claim of the theorem is established. □

Corollary. The generator polynomial of a cyclic $[n, k]$ code has degree $n - k$.

We still have no method of actually finding cyclic codes. Here is a theorem which helps considerably.

Theorem 8.5 The generator of any ideal of $F[X]/x^n - 1$ is a divisor of $x^n - 1$.

Proof. Let g be the generator of ideal I. Divide $x^n - 1$ by $g(x)$ to get $x^n - 1 = g(x)q(x) + r(x)$ with $\deg(r) < \deg(g)$.
Then $r(x) \equiv g(x)q(x)$ so $r \in I$.
Hence $r = 0$ and $x^n - 1 = g(x)q(x)$. □

[Ex 5–7]

From the results developed so far it looks as if finding all cyclic codes of length n over F is equivalent to the problem of finding all cyclic code generators, which in turn is equivalent to finding all monic divisors of $x^n - 1$. But to make this work, suppose $g(x)|x^n - 1$. Does it follow that $g(x)$ is *the* generator of $\langle g(x) \rangle$, or could there be a polynomial $a(x)g(x)$ which has degree smaller than $\deg(g)$ when reduced modulo $x^n - 1$?

Theorem 8.6 Let I be the ideal $\langle g(x) \rangle$ in $F[X]/x^n - 1$, where g is monic and $g(x)|x^n - 1$. Let $h(x)$ be *the* generator of I. Then $g = h$.

Proof. $h \in I$ so for some polynomial $a(x), h(x) \equiv a(x)g(x)$.
That is $h(x) = a(x)g(x) + b(x)(x^n - 1)$. Since $g(x)|x^n - 1, g(x)$ divides the right hand side of this equation. Hence $g|h$. But $\deg(h) < \deg(g)$, and both g, h are monic so $g = h$ □

From these results we see that each ideal of $f[X]/x^n - 1$ corresponds to a unique divisor of $x^n - 1$, its generator, so the number of cyclic codes of length n over F is the number of monic divisors of $x^n - 1$ so we now need a method of enumerating these divisors.

Theorem 8.7 If the unique factorization of $x^n - 1$ into monic irreducibles is $x^n - 1 = (f_1(x))^{t_1} \cdots (f_m(x))^{t_m}$, then there are $(1+t_1)\cdots(1+t_m)$ monic divisors of $x^n - 1$, each being of the form

$$(f_1(x))^{u_1} \cdots (f_m(x))^{u_m} \text{ where } 0 \le u_i \le t_i \text{ for } 1 \le i \le m.$$

Proof. The form of the divisors follows from the unique factorization theorem for $f[X]$. The rest is an easy counting argument. □

Factorizing $x^n - 1$ completely is not easy, and several ingenious methods and many tables have been produced. The topic will not be pursued in this book, but we give a reference table below for $n \leq 25$ over the binary field.

The factorization of $x^n - 1$ into irreducibles over Z_2. [Taken from [17]]

n	factorization
1	$1+x$
2	$(1+x)^2$
3	$(1+x)(1+x+x^2)$
4	$(1+x)^4$
5	$(1+x)(1+x+x^2+x^3+x^4)$
6	$(1+x)^2(1+x+x^2)^2$
7	$(1+x)(1+x+x^3)(1+x^2+x^3)$
8	$(1+x)^8$
9	$(1+x)(1+x+x^2)(1+x^3+x^6)$
10	$(1+x)^2(1+x+x^2+x^3+x^4)^2$
11	$(1+x)(1+x+\cdots+x^{10})$
12	$(1+x)^4(1+x+x^2)^4$
13	$(1+x)(1+x+\cdots+x^{12})$
14	$(1+x)^2(1+x+x^3)^2(1+x^2+x^3)^2$
15	$(1+x)(1+x+x^2)(1+x+x^2+x^3+x^4)(1+x+x^4)(1+x^3+x^4)$
16	$(1+x)^{16}$
17	$(1+x)(1+x+x^2+x^4+x^6+x^7+x^8)(1+x^3+x^4+x^5+x^8)$
18	$(1+x)^2(1+x+x^2)^2(1+x^3+x^6)^2$
19	$(1+x)(1+x+x^2+\cdots+x^{18})$
20	$(1+x)^4(1+x+x^2+x^3+x^4)^4$
21	$(1+x)(1+x+x^2)(1+x^2+x^3)(1+x+x^3)(1+x^2+x^4+x^5+x^6)$ $\times(1+x+x^2+x^4+x^6)$
22	$(1+x)^2(1+x+x^2+\cdots+x^{10})^2$
23	$(1+x)(1+x+x^5+x^6+x^7+x^9+x^{11})(1+x^2+x^4+x^5+x^6+x^{10}+x^{11})$
24	$(1+x)^8(1+x+x^2)^8$
25	$(1+x)(1+x+x^2+x^3+x^4)(1+x^5+x^{10}+x^{15}+x^{20})$.

For example $Z_2[X]/x^9-1$ has $2^3 = 8$ ideals (cyclic codes) consisting of those with generator polynomials of the form $(1+x)^a(1+x+x^2)^b(1+x^3+x^6)^c$ with a, b, c each equal to 0 or 1. Their degrees are 0, 1, 2, 3, 6, 7, 8 or 9, so the possible dimensions are 9, 8, 7, 6, 3, 2, 1.

We can also use these results to find the smallest cyclic code containing a given word. For example, if the cyclic code C over Z_2 contains the word 101001001, what are the possibilities? This word is represented by the polynomial $1+x^2+x^5+x^8$. Its factorization into irreducibles over Z_2 is $(1+x)(1+x+x^5+x^6+x^7)$ (check this). The generator polynomial of C is of the form given above, so our given word (polynomial) must be a multiple of this modulo $x^9 - 1$. So

$$\begin{aligned}(x+1)(x^7+x^6+x^5+x+1) &= \lambda(x)(1+x)^a(1+x+x^2)^b \\ &\quad \times(1+x^3+x^6)^c + \mu(x)(x^9-1)\end{aligned}$$

If $b > 0$ then $1 + x + x^2$ divides the right hand side of this. But check that it does not divide the left hand side, so $b = 0$. Similarly $c = 0$. Hence the generator of C can only be 1 or $1 + x$. The latter gives the smallest code. [Ex 8, 9]

8.4 Encoding by polynomials

The generator matrix we have produced does not have the nice 'nearly standard form' of Chapter 5. Our first job in this section is to manufacture such a form for cyclic codes. The result of encoding the message $\boldsymbol{m}$ as the codeword $\boldsymbol{m}G = \boldsymbol{c}$ when G is 'nearly standard' is that $\boldsymbol{c}$ will have the message symbols in those places corresponding to the columns of G which make up the unit matrix.

Let $g(x)$ be the generator polynomial of the $[n, k]$ cyclic code C. Divide each of the powers x^{n-k+i} for $i = 0, 1, 2, \cdots, k-1$ by $g(x)$ to obtain $x^{n-k+i} = q_i(x)g(x) + r_i(x)$, where, as usual, $\deg(r_i) < \deg(g) = n - k$.

$x^{n-k+i} - r_i(x) = q_i(x)g(x)$, which is a codeword since $C = \langle g(x) \rangle$. Hence the matrix G whose rows are $x^{n-k+i} - r_i(x)$ is a generator matrix for C because it has the right number of rows ($k = \dim C$), each row is a codeword, and the rows are independent because the last k columns make I_k. To see this, if row i is $c_{i0}c_{i1}c_{i2} \cdots c_{i\,n-1}$, then $c_{i0}, c_{i1}, \cdots, c_{i\,n-k-1}$ are just the coefficients of $-r_i(x)$, and $c_{i\,n-k}, \cdots, c_{i\,n-1}$ are all zero except for $c_{i\,n-k+i} = 1$, corresponding to x^{n-k+i}. [Ex 10]

Theorem 8.8 Let G be the nearly standard generator matrix for the $[n, k]$ cyclic code C, as described above. Then the codeword $\boldsymbol{c}$ arising from the message $\boldsymbol{m}$ has polynomial representation $c(x) = q(x)g(x)$ where $q(x)$ is the quotient on dividing $x^{n-k}m(x)$ by $g(x)$.

$$\text{i.e.} x^{n-k}m(x) = q(x)g(x) + r(x) \text{with } \deg(r) < \deg(g) = n - k.$$

Proof. Let $m(x) = m_0 + m_1x + \cdots + m_{k-1}x^{k-1}$. Multiplying the equations in the preamble to the theorem by $m_0, m_1, \cdots, m_{k-1}$ respectively we have

$$\begin{array}{lcll}
m_0x^{n-k} & = & m_0\,q_0(x)g(x) & + \quad m_0\,r_0(x) \\
m_1x\,x^{n-k} & = & m_1\,q_1(x)\,g(x) & + \quad m_1\,r_1(x) \\
 & \vdots & & \\
 & \vdots & & \\
m_{k-1}x^{k-1}\,x^{n-k} & = & m_{k-1}\,q_{k-1}(x)g(x) & + \quad m_{k-1}\,r_{k-1}(x)
\end{array}$$

Summing these:

$$\begin{aligned}
m(x)x^{n-k} \;=\; & g(x)\left[m_0\,q_0(x) + m_1\,q_1(x) + \cdots + m_{k-1}q_{k-1}(x)\right] \\
& + \left[m_0\,r_0(x) + m_1\,r_1(x) + \cdots + m_{k-1}\,r_{k-1}(x)\right]
\end{aligned}$$

Now the second $[\cdots]$ term has degree $< n-k$, so by the uniqueness clause in the division algorithm, this term is $r(x)$ and the other $[\cdots]$ term is $q(x)$.

Also $g(x)[m_0q_0(x) + m_1q_1(x) + \cdots + m_{k-1}q_{k-1}(x)]$ is the linear combination of rows of G : m_0 row $0 + m_1$ row $1 + \cdots + m_{k-1}$ row $k-1$, which is $\boldsymbol{m}G = \boldsymbol{c}$. □

[Ex 11]

The point of this reformulation of the encoding process in terms of polynomial operations rather than matrix operations is that the polynomial operations can be done very rapidly by simple bits of hardware called feed back shift registers, and the algorithms on which they operate only require storage of the generator polynomial, not the generator matrix. We shall not pursue this topic further, but a good account appears in [12].

8.5 Syndromes and polynomials

Having obtained a nearly standard generator for a cyclic code we can use the technique in section 5.7 to obtain its parity check matrix. From this we may calculate syndromes, give their polynomial interpretation, and find a useful decoding algorithm.

The generator matrix constructed in section 8.4 has the form $G = [R|I]$ where the rows of R are $-r_0(x), -r_1(x), \cdots, -r_{k-1}(x)$ and I is the $k \times k$ unit matrix. Hence a parity check matrix for C is $H = [J| - R^t]$ where J is the $(n-k) \times (n-k)$ unit matrix and R^T is the transpose of R. The advantage of this choice of parity check matrix for C is that the following neat result holds for it.

Theorem 8.9 Let $a(x)$ be a polynomial of degree at most $n-1$ corresponding to the word $\boldsymbol{a}$ arriving at the decoder, and let $s(x)$ be the polynomial of its syndrome $\boldsymbol{s}$. Then when $a(x)$ is divided by $g(x)$ the remainder is $s(x)$.

Proof. $\boldsymbol{s} = \boldsymbol{a}H^T = a_0$ (column 0 of H) $+ \cdots + a_{n-1}$(column $n-1$ of H).

So $s(x) = a_0 1 + a_1x + a_2x^2 + \cdots + a_{n-k-1}x^{n-k-1}$

$$+a_{n-k}r_0(x) + a_{n-k+1}r_1(x) + \cdots + a_{n-1}r_{k-1}(x), \qquad (1)$$

(replacing the columns of H by their polynomial representatives)

$$= a_0 + a_1x + \cdots + a_{n-k-1}x^{n-k-1} + a_{n-k}(x^{n-k} - q_0(x)g(x))$$
$$+ \cdots + a_{n-1}(x^{n-1} - q_{k-1}(x)g(x))$$
$$= a(x) - g(x)[a_{n-k}q_0(x) + a_{n-k+1}q_1(x) + \cdots + a_{n-1}q_{k-1}(x)]$$

which we write as $s(x) = a(x) - g(x)Q(x)$. (2)

But from equation (1) above $s(x)$ has degree $\leq n-k-1$, so it follows from (2) that $s(x)$ is the remainder on dividing $a(x)$ by $g(x)$. □

[Ex 12]

Recalling that in $F[X]/x^n - 1$ multiplication by x corresponds to cyclically shifting a word, we can easily derive from the previous theorem a simple connection between the syndromes of $\boldsymbol{a}$ and $\boldsymbol{a}^1$.

$$\begin{array}{lrcl} \text{From that theorem,} & a(x) & = & q(x)g(x) + s(x) \\ \text{So} & x\,a(x) & = & xq(x)g(x) + xs(x) \\ & & = & xq(x)g(x) + Q(x)g(x) + t(x) \end{array}$$

where $Q(x)$, $t(x)$ are the quotient and remainder when $x\,s(x)$ is divided by $g(x)$. Now deg $(s) \leq n-k-1$ and $g(x)$ is monic and has degree $n-k$. So if deg$(s) < n-k-1$, $Q(x) = 0$ and $t(x) = xs(x)$. When deg$(s) = n-k-1$ $Q(x)$ is the constant s_{n-k-1} and $t(x) = x\,s(x) - s_{n-k-1}g(x)$. These results are covered by the following theorem.

Theorem 8.10 For the cyclic $[n,k]$ code C with generator $\boldsymbol{g}$ and parity check matrix as defined in this section, if syn$(\boldsymbol{a}) = \boldsymbol{s}$ then syn$(\boldsymbol{a}^1)$ has representative polynomial $x\,s(x) - s_{n-k-1}g(x)$. □

[Ex 13]

A natural investigation now is to find a polynomial version of syndrome decoding. Let C be an $[n,k]$ code, not necessarily cyclic, and let $d(C) = 2t+1$ so that C is t-error correcting. We also assume that C has a parity check matrix of the form $H = [J|A]$ where J is the $(n-k)\times(n-k)$ unit matrix. Let $\boldsymbol{l}$ be an error pattern of weight $\leq t$ whose syndrome $\boldsymbol{s}$ also has weight $\leq t$. $\boldsymbol{l}$ is therefore a correctable error, so it is a coset leader in the Slepian array. Define the n symbol word $\boldsymbol{s}^*$ with $s_0s_1\cdots s_{n-k-1}$ as its initial segment and zeros in the remaining positions, so $\boldsymbol{s}^* = \boldsymbol{s}|\boldsymbol{0}$. Then clearly $\boldsymbol{l}H^T = \boldsymbol{s} = \boldsymbol{s}^*H^T$, so $\boldsymbol{l}$ and $\boldsymbol{s}^*$ have the same syndrome. They are therefore in the same row (coset) of the array, and since their weights are both $\leq t$ they are both equal to the coset leader of that row. That is, $\boldsymbol{l} = \boldsymbol{s}^*$. In those cases where the syndrome of the received word has weight $\leq t$, this result means that the error pattern can be deduced immediately – no searching of the coset leader/syndrome list is necessary, which for large codes can mean a significant saving of time. The snag of course is that in general the syndrome weight will not be $\leq t$. Nevertheless, if we now specialize to *cyclic* codes we can derive a useful error-correcting technique sometimes called *error-trapping.*

Let C be cyclic $[n,k]$ with $d(C) = 2t+1$, and parity check matrix as above. Suppose a transmitted word is corrupted by error $\boldsymbol{e}$ (with weight $\leq t$) , and that $\boldsymbol{e}$ has a continuous run (counted cyclically) of at least k zeros. (For example $\underline{00}101101\underline{00}$ and $11\underline{0000}1010$ both have cyclic runs of 4 zeros). Let $\boldsymbol{c}$ and $\boldsymbol{r}$ be the transmitted and received words respectively. Now for some i, $\boldsymbol{e}^i$ has 0 in its last k places, and $w(\boldsymbol{e}^i) \leq t$. Hence $\boldsymbol{e}^i = \boldsymbol{f}|\boldsymbol{0}$, using $\boldsymbol{0}$ here to denote the zero word of length k, so syn$(\boldsymbol{e}^i) = \boldsymbol{f}$, and $w(\boldsymbol{f}) \leq t$.

Now let us see how the receiver can exploit the previous two theorems and the discussion above to obtain a decoding strategy. $\boldsymbol{r}$ is known so syn($\boldsymbol{e}$) is known as syn($\boldsymbol{r}$) = syn($\boldsymbol{e}$). syn($\boldsymbol{e}^i$) can then be calculated for all i using Theorem 8.10, and the smallest i for which syn($\boldsymbol{e}^i$) has weight $\leq t$ is recorded. Then $\boldsymbol{e}^i$ is known to be syn($\boldsymbol{e}^i$) followed by k zeros, so by cycling this back i places, $\boldsymbol{e}$, and hence $\boldsymbol{c}$ is recovered. By using Theorem 8.9 too, all the calculations can be carried out in terms of polynomials. Here is an example.

Let C be the binary [15, 7] cyclic code generated by $g(x) = 1 + x^4 + x^6 + x^7 + x^8$. This code in fact has $d(C) = 5$ so it is 2-error correcting. Suppose 111101010010010 is received. We carry out the decoding procedure outlined above. The received word has polynomial $r(x) = 1 + x + x^2 + x^3 + x^5 + x^7 + x^{10} + x^{13}$. Its syndrome $s(x)$ is the remainder on dividing $r(x)$ by $g(x)$, and you should check that $s(x) = 1 + x^2 + x^3 + x^4 + x^5$, and then check that, using the method of Exercise 13, the first cyclic shift of the error vector which has a syndrome weight $\leq$ 2 is $\boldsymbol{e}^{10}$, and syn($\boldsymbol{e}^{10}$) = 10000010, so $\boldsymbol{e}^{10}$ = 100000100000000. $\boldsymbol{e} = \boldsymbol{e}^{15}$ so $\boldsymbol{e}$ is obtained by cycling $\boldsymbol{e}^{10}$ a further 5 places : $\boldsymbol{e}$ = 000001000001000. Hence the transmitted word $\boldsymbol{c} = \boldsymbol{r} - \boldsymbol{e}$ = 111100010011010. You can check that this is indeed a codeword by dividing its polynomial by $g(x)$ and obtaining a remainder of zero.

Notice that the success of the method depends on $\boldsymbol{e}$ having a cyclic run of at least k zeros. In our example above $n = 15$ and $k = 7$ so *all* errors of weight $\leq$ 2 will satisfy this condition. [Ex 14]

8.6 Parity checks and polynomials

In this section we give a polynomial interpretation of the parity check matrix of a cyclic code. Let $C = \langle g(x) \rangle$ be a cyclic $[n, k]$ code. Hence $g(x) | x^n - 1$ so $g(x)h(x) = x^n - 1$ for some h. Now $\deg(g) = n - k$ so $\deg(h) = k$. Also $g(x)$ and $x^n - 1$ are monic, so h is monic.

Definition 8.2 The polynomial h introduced above is called the check polynomial of C.

In spite of its name h does not in general generate $C^\perp$ but there is a close connection between h and $C^\perp$ as we shall see shortly.

Theorem 8.11 $c(x)$ corresponds to a codeword of C if and only if $c(x)h(x) \equiv 0 \bmod x^n - 1$.

Proof.

$$\begin{aligned} \boldsymbol{c} \in C \quad \Rightarrow \quad c(x) &\equiv a(x)g(x) \bmod x^n - 1 \text{ for some } a(x) \in F[X] \\ \Rightarrow \quad c(x)h(x) &\equiv a(x)g(x)h(x) \equiv 0 \bmod x^n - 1 \end{aligned}$$

Conversely, suppose $c(x)h(x) \equiv 0$. Divide $c(x)$ by $g(x)$ to get $c(x) = g(x)q(x) + r(x)$ with deg $(r) < n - k$.

$$\begin{aligned} \text{Then } c(x)h(x) \equiv 0 \quad &\Rightarrow \quad g(x)q(x)h(x) + r(x)h(x) \equiv 0 \\ &\Rightarrow \quad r(x)h(x) \equiv 0 \\ &\Rightarrow \quad r(x)h(x) \text{ is a multiple of } x^n - 1 \end{aligned}$$

But $\deg(r) < n - k$ and $\deg(h) = k$, so $\deg(rh) < n$, and hence $rh = 0$.
Therefore $r = 0$ so $c(x) = g(x)q(x)$; that is, $\boldsymbol{c} \in C$. □

We have seen (Theorem 6.19) that the duals of cyclic codes are themselves cyclic. It remains to find the generator polynomial of the dual, which will then (via Theorem 8.2) give us a generator matrix for $C^\perp$, in other words a parity check matrix for C. In passing we make use of another part of Theorem 8.2 – that if $g(x)$ is the generator of C, then $g_0 \neq 0$. Now suppose $p(x)$ is any polynomial of degree m with a non-zero constant term. It is easy to check that if we regard 'x' as a formal object which we manipulate as in ordinary algebra, then $x^m p(x^{-1})$ is also a polynomial of degree m with non-zero constant term, obtained by reversing the order of the sequence of coefficients in p. [Ex 15]

Theorem 8.12 If $h(x)$ is the check polynomial of the $[n,k]$ cyclic code C, then $\overline{h}(x) = h_0^{-1}x^k h(x^{-1})$ is the generator polynomial of $C^\perp$

Proof. $\overline{h}(x)$ is clearly monic of degree k, and by considering the constant term in the relation $g(x)h(x) = x^n - 1$ we see that $h_0 \neq 0$.

$g(x)h(x) = x^n - 1 \Rightarrow g(x^{-1})h(x^{-1}) = (x^{-1})^n - 1 = x^{-n} - 1$, and this can be written $-x^n h_0\ h_0^{-1}h(x^{-1})g(x^{-1}) = x^n - 1$. i.e. $-h_0 g(x^{-1})x^{n-k}\overline{h}(x) = x^n - 1$.

Now $x^{n-k}g(x^{-1})$ is a polynomial of degree $n-k$ since $g_0 \neq 0$.

Hence $\overline{h}(x)$ is a monic divisor of $x^n - 1$, with degree k, so $\langle \overline{h} \rangle$ is an $[n,\ n-k]$ cyclic code. But $\dim(C^\perp) = n - k$ so we now show that $\langle \overline{h} \rangle \subseteq C^\perp$ and deduce, from item 11(b) of section 5.3, that $\langle \overline{h} \rangle = C^\perp$.

$\overline{h}(x), g(x)$ represent the words $h_k h_{k-1} \cdots h_0 0 0 \cdots 0$ and $g_0 g_1 \cdots g_{n-k} 0 0 \cdots 0$ respectively. These are orthogonal because their dot product, $g_0 h_k + g_1 h_{k-1} + \cdots$ is simply the coefficient of x^k in $h(x)g(x) = x^n - 1$ $(0 < k < n)$. Similarly the dot products of $\overline{\boldsymbol{h}}$ with $\boldsymbol{g}^1, \boldsymbol{g}^2, \cdots, \boldsymbol{g}^{k-1}$ are the coefficients (in $x^n - 1$) of $x^{k-1}, x^{k-2}, \cdots, x$ respectively, all of which are zero. Hence $\overline{\boldsymbol{h}} \in C^\perp$ because it is orthogonal to every row of the generator matrix for C given by Theorem 8.2.

But $C^\perp$ is cyclic, so every cyclic shift of $\overline{\boldsymbol{h}}$ is in $C^\perp$. Every polynomial multiple of $\overline{h}(x)$ corresponds to a linear combination of cyclic shifts of $\overline{\boldsymbol{h}}$, so by linearity these are all in $C^\perp$ too. That is $\langle \overline{h}(x) \rangle \subseteq C$, as claimed. □

Corollary. If $h(x) = h_0 + h_1 x + \cdots + h_k x^k$ is the check polynomial of C, then the matrix

$$H = \begin{bmatrix} h_k & h_{k-1} & \cdots & \cdots & h_0 & 0 & 0 & \cdots & 0 \\ 0 & h_k & \cdots & \cdots & h_1 & h_0 & 0 & \cdots & 0 \\ & & & & \vdots & & & & \\ & & & & \vdots & & & & \\ 0 & 0 & \cdots & h_k & \cdots & \cdots & \cdots & \cdots & h_0 \end{bmatrix}$$

is a generator matrix for $C^{\perp}$, that is, a parity check matrix for C. □

[Ex 16, 17]

8.7 Cyclic codes and double-adjacent errors

Throughout this book we have taken the view that errors occur at random in a word so that it was reasonable to measure the error correcting performance of a code in terms of the *number* of correctable error patterns of each weight. We effectively ignored the *distribution* of errors within the word. An important practical application of coding theory is to channels which induce errors in short bursts rather than scattered through the word, and the techniques studied in this chapter can be used to devise good 'burst error correcting' cyclic codes. We refer you to other sources such as [17] for further details of both the theory and application. In this section we look at the simple case of 'double adjacent' errors – that is error patterns of weight 2 in which the two non-zero symbols are adjacent (including the case of one at each end of the word since we continue to think cyclically). To be specific, we seek a cyclic code which will correct (at least) all single errors and all double adjacent errors. The main result for binary codes is

Theorem 8.13 If a binary cyclic $[n,k]$ code C has generator $g(x) = (x+1)p(x)$ with $p(x) \not| x^i - 1$ for $i = 1, 2, \cdots, n-1$, then C will correct all single and all double adjacent errors.

Proof. From Exercise 6b the fact that $x+1|g(x)$ implies that all words of C have even weight. Hence no word of weight one can be a codeword, so all such words appear in any Slepian array for C somewhere other than the first row. Furthermore no two words of weight one can appear in the same row, since it is easy to derive a contradiction from words with polynomial representations x^a and $x^b (0 \le a < b < n)$ in the same row: it would mean x^a and x^b had the same syndrome, which in turn means their remainders on division by $g(x)$ are the same, so their difference is a multiple of $g(x)$. But $g(x)|x^b - x^a \Rightarrow g(x)|x^a(x^{b-a} - 1) \Rightarrow g(x)|x^{b-a} - 1$ since $gcd(g(x), x^a) = 1, \Rightarrow p(x)|x^{b-a} - 1$ which contradicts the hypothesis of the theorem. Hence the words of weight one appear in distinct rows, with none in the first, so they can all be chosen as their coset leaders, so all weight one errors are correctable. Hence $d(C) \ge 3$, and since C is an even weight code $d(C) \ge 4$.

This implies that no weight 1 and weight 2 words can be in the same coset because their difference would be a codeword of weight at most 3.

Finally we show that no pair of weight 2 'double adjacent' words can appear in the same coset. Assuming the contrary, suppose $x^a + x^{a+1}$ and $x^b + x^{b+1}$ were in the same row. These four terms must be distinct since otherwise their difference would have weight ≤ 2. The difference is the codeword $x^a + x^{a+1} + x^b + x^{b+1} = x^a(1+x)(1+x^{b-a})$, a multiple of $g(x)$.

But $\gcd(x^a,\ g(x)) = 1$ so $g(x)|(1+x)(1-x^{b-a})$, and hence $p(x)|1-x^{b-a}$, another contradiction.

The outcome of all this is that $\mathbf{0}$, all words of weight 1 and all double adjacent words of weight 2 appear in distinct cosets, so they are all minimal weight words of their respective cosets and can therefore be chosen as coset leaders. □

[Ex 18]

8.8 Cyclic golay codes

In Chapter 6 we found that the only perfect non-trivial linear codes were the Hamming codes, binary codes with parameters [23, 12, 7] and ternary codes with parameters [11, 6, 5]. We also stated (but did not prove) that the latter two codes were unique up to equivalence, and gave a rather ad hoc construction of the binary one. In this section both are presented as cyclic codes, from which an alternative deduction of their minimum distances is given. The method is essentially that of Hill [1], in which an ingenious chain of results leads eventually to the minimum distance result. Some of the steps are left as exercises, and some of these involve the congruence results of Chapter 3. We start with a definition.

Definition 8.3 let f be a polynomial of degree n. The reversal of f, which we denote by f^{rev}, is defined as the polynomial whose coefficient sequence is the reverse of that of f.

We shall be mainly concerned with a polynomial g which is the generator of a cyclic code, so that it has a non-zero constant term. In this case g^{rev} has the same degree as g. For words $\boldsymbol{a}$ rather than polynomials $a(x)$ we shall write the word which is $\boldsymbol{a}$ with its digits written in reverse order as $\text{Rev}(\boldsymbol{a})$.

The particular feature which is the key to this treatment of the perfect Golay codes is that $x^{23}-1$ over Z_2 and $x^{11}-1$ over Z_3 have factorizations of the form $(x-1)g(x)g^{rev}(x)$ and $-(x-1)g(x)g^{rev}(x)$ respectively. Note that for words with non-zero constant terms $g^{rev}(x) = g_0\,\overline{g}(x)$, so in the binary case $g^{rev}(x) = \overline{g}(x)$. [Ex 19]

First we tackle the binary code

Theorem 8.14 Let p be an odd prime for which $x^p - 1$ factorizes over Z_2 as $(x-1)g(x)g^{rev}(x)$. Then all codewords of $\langle g\rangle$ with weight w satisfy $w^2 - w \geq p-1$ if w is odd, and $w \equiv 0 \bmod 4$ if w is even, with $w \neq 4$ unless $p = 7$.

Proof. By Exercise 19 g and g^{rev} generate equivalent codes and we saw in Chapter 4 that equivalent codes have identical distance distributions. For *linear* codes this means they have identical weight distributions, so for each word $\boldsymbol{c}$ of weight w in $\langle g\rangle$ there is a corresponding word $\boldsymbol{c}'$ in $\langle g^{rev}\rangle$ with the same weight. From the solution to the same exercise we see that $\boldsymbol{r}$ is a row of G if and only if Rev($\boldsymbol{r}$) is a row of G'. But 'Rev' is a linear operator (Rev($\lambda\boldsymbol{a} + \mu\boldsymbol{b}$) $= \lambda$ Rev($\boldsymbol{a}$) $+ \mu$ Rev($\boldsymbol{b}$)), so it follows that the words of $\langle g\rangle$ are just the reversals of the words of $\langle g^{rev}\rangle$.

Now $g(x)g^{rev}(x) = \dfrac{x^p - 1}{x - 1} = 1+x+x^2+\cdots+x^{p-1}$, which is a word of odd weight p. Let $\mathbf{c} \in \langle g\rangle$ with odd weight w, then $\mathbf{c}' = \text{Rev}\ (\mathbf{c})$ has the same weight and is a member of $\langle g^{rev}\rangle$. The polynomials corresponding to $\mathbf{c}$ and $\mathbf{c}'$ are multiples of g and g^{rev} respectively, so their product is a multiple of $1+x+x^2+\cdots+x^{p-1}$. Working modulo x^p-1, a multiple of a polynomial f is just a sum (mod 2) of cyclic shifts of f, so $c(x)c'(x) \equiv 1+x+x^2+\cdots+x^{p-1}$ or 0 mod x^p-1, depending on whether the sum has an odd or even number of terms. As polynomials in $Z_2[X]/x^p-1$ $c(x)$, $c'(x)$ both have w(odd) non-zero terms. Hence $c(1)c'(1) = w^2 \equiv 1$, so $c(x)c'(x)$ cannot be a multiple of $x^p - 1$. So

$$c(x)c'(x) \equiv 1 + x + x^2 + \cdots + x^{p-1} \bmod x^p - 1 \qquad (1)$$

Now $c(x)c'(x) = (c_0+c_1\,x+\cdots c_{p-1}x^{p-1})(c_{p-1}+c_{p-2}\,x+\cdots+c_0\,x^{p-1})$, where exactly w of $c_0, c_1, \cdots, c_{p-1}$ are non-zero.

Multiplying out the left hand side of (1), w of the w^2 non-zero terms are of the form $c_i^2x^{p-1}$, so these must sum to the single term x^{p-1} on the right hand side. The remaining $w^2 - w$ non-zero terms on the left must account for the other $p-1$ non-zero terms on the right, so $w^2 - w \geq p-1$. [Note that $\geq$ rather than $=$ is correct here because each x^u on the right is the sum of one *or more* x^u terms on the left.] This completes the proof for codewords of odd weight.

Now take $\mathbf{c}$, $\mathbf{c}'$ =Rev ($\mathbf{c}$) to be codewords of *even* weight w in $\langle g\rangle$, $\langle g^{rev}\rangle$ respectively. An identical argument to that already used leads this time to

$$c(x)c'(x) \equiv 0 \bmod x^p - 1 \qquad (2)$$

$c(x)$ can be written $x^{e_1}+x^{e_2}+\cdots+x^{e_w}$ where only those terms corresponding to the non-zero bits of $\mathbf{c}$ have been written down, and $0 \leq e_i < p$. Then $c'(x) = x^{p-1}c(x^{-1}) = x^{p-1-e_1} + x^{p-1-e_2} + \cdots + x^{p-1-e_w}$, and from (2) above, when we use these expressions to expand $c(x)c'(x)$ as a polynomial every power of x has an even coefficient, that is, 0 mod 2, and of course $x^{t\pm np}$ is counted as x^t.

In the expansion the term x^{p-1} occurs an even number of times because it can only arise from products of the form $x^{e_i}x^{p-1-e_i}$ and there are w of these. Now consider the remaining $w^2 - w$ terms. A typical one is $x^{p-1-e_j+e_i}$ with $i \neq j$. This comes from $x^{e_i}x^{p-1-e_j}$, and if the total number of occurrences of this term is even it must occur again, say as $x^{e_k}x^{p-1-e_l}$, with $i \neq k$, $j \neq l$ and $l \neq k$. Hence $e_i - e_j \equiv e_k - e_l \bmod p$, but this implies $e_i - e_k \equiv e_j - e_l \bmod p$, so we have another pair of cancelling terms $x^{e_i}x^{p-1-e_k}$ and $x^{e_j}x^{p-1-e_l}$. Hence $w^2 - w$ is a multiple of 4. But $w(w-1) \equiv 0 \bmod 4$ implies $w \equiv 0 \bmod 4$ since $w-1$ is odd so gcd(4, $w-1) = 1$. [Ex 20]

Finally, suppose $w = 4$, and we then write $c(x)$ as $x^a + x^b + x^c + x^d$. Hence $c(x)c'(x) = (x^a+x^b+x^c+x^d)(x^{p-1-a}+x^{p-1-b}+x^{p-1-c}+x^{p-1-d})$. Expanding $c(x)c'(x) \equiv 0 \bmod x^p - 1$, we have 4 terms x^{p-1} on the left hand side, and the remaining 12 terms are all those of the form $x^{i+p-1-j}$ where i, j are distinct members of $\{a, b, c, d\}$. The cancelling pairs of terms are those whose powers are congruent mod p. So ignoring the $p-1$ which is common to all 12 powers, the twelve split into six pairs of the form $(i-j, j-i)$ which cannot be congruent mod p as this would imply $2(i-j) \equiv 0 \bmod p$ which is impossible since p is odd and i, j are distinct and between 0 and $p-1$ inclusive. For similar reasons $(i-j, i-k)$ cannot be a cancelling pair. Another case we can quickly exclude is the pair of terms $x^{i+p-1-j}$, $x^{k+p-1-l}$ where i, j, k, l are distinct, for suppose $i - j \equiv k - l$. This forces $i - k \equiv j - l$, and of course $j - i \equiv l - k$ and $k - i \equiv l - j$. Now what can cancel with $i - l$? Only $j-k$, $k-j$, $l-i$ are left, and as we have already remarked, the last of these is not possible. Hence we are reduced to

(1)	$i-j \equiv k-l$		(1)	$i-j \equiv k-l$
(2)	$i-k \equiv j-l$	or	(2)	$i-k \equiv j-l$
(3)	$i-l \equiv j-k$		(3)	$i-l \equiv k-j$

But the second set of congruences is not a separate case to consider because it is just the result of interchanging j and k in the first set. So working with the first set we derive : $2i - j - l \equiv j - l$ adding (1) and (3).

$\Rightarrow 2(i-j) \equiv 0$ which is impossible.

Hence we must have a cancelling pair of the form $i - j \equiv j - k$ where i, j, k are distinct, with no cancelling pair of the form just excluded. This leads to

(1)	$i-j \equiv j-k$		(1)	$i-j \equiv j-k$
(2)	$i-k \equiv k-l$	or	(2)	$i-k \equiv l-i$
(3)	$i-l \equiv l-j$		(3)	$j-l \equiv l-k$

and interchanging i and k in the first set yields the second set, so only the first needs to be examined. Using (1) and (2) to write i and l in terms of j and k, $i = 2j - k$ and $l = 3k - 2j$. Substituting these in (3) we obtain $7(j-k) \equiv 0 \bmod p$ which implies $p = 7$ since $j - k \not\equiv 0 \bmod p$. □

Corollary. The code $\langle g \rangle$ investigated above is equivalent to the binary [23, 12, 7] Golay code of Chapter 6. [Ex 21]

In the next theorem we develop a sequence of results culminating in a proof that a certain cyclic code over Z_3 has all the parameters of the perfect ternary Golay code, and is therefore equivalent to that code.

Over Z_3 you may check that

$$\begin{aligned} x^{11}-1 &= -(x-1)(x^5+x^4-x^3+x^2-1)(-x^5+x^3-x^2+x+1) \\ &= -(x-1)g(x)g^{rev}(x). \end{aligned}$$

So the cyclic codes $\langle g \rangle$ and $\langle g^{rev} \rangle$ are equivalent and have dimension 6 and length 11. Let C and D be the cyclic codes $\langle g(x) \rangle$ and $\langle (x-1)g(x) \rangle$ respectively, and let $\mathbf{c}$ be a codeword of C with weight w. In the statement of our theorem all numerical congruences are modulo 3.

Theorem 8.15

(i) $\mathbf{c} \in D$ if and only if $\sum_{i=0}^{10} c_i \equiv 0$;

(ii) if $\mathbf{c} \in D$ then $w \equiv 0$;

(iii) if $\mathbf{c} \notin D$ then $w \equiv 2$, and $w \geq 5$;

(iv) $w \neq 3$;

(v) $d(C) = 5$.

Proof.

(i) Note that if $\mathbf{c}$ has polynomial $c(x)$, then $c(1) = c_0 + c_1 + \cdots + c_{10}$.
Hence $\mathbf{c} \in D \Rightarrow x-1 | c(x) \Rightarrow c(1) \equiv 0 \Rightarrow \sum_{i=0}^{10} c_i \equiv 0$.

For the converse, $\sum_{i=0}^{10} c_i \equiv 0 \Rightarrow c(1) \equiv 0 \Rightarrow x-1 | c(x)$.

But $\mathbf{c} \in C$ so $g(x)|c(x)$, and $x-1 \not| g(x)$ so $gcd(x-1, g(x)) = 1$.
Hence $g(x)(x-1)|c(x)$ and $\mathbf{c} \in D$.

(ii) The check polynomial of D is $-g^{rev}(x)$, so the generator polynomial of $D^{\perp}$, by Theorem 8.12, is

$$-x^5(x^{-5}-x^{-3}+x^{-2}-x^{-1}-1) = -1+x^2-x^3+x^4+x^5 = g(x).$$

Hence $D^{\perp} = C$, but $D \subseteq C$ because $(x-1)g(x)|c(x) \Rightarrow g(x)|c(x)$, so $D \subseteq D^{\perp}$.

Thus D is self-orthogonal and any two codewords of D have zero dot product. In particular, for each $\mathbf{c} \in D, \mathbf{c} \cdot \mathbf{c} = 0$, in other words $\sum_{i=0}^{10} c_i^2 \equiv 0$. But for $c_i \neq 0, c_i^2 \equiv 1$, so this sum is just the weight of $\mathbf{c}$ and we have $w \equiv 0$. [Ex 22]

(iii) Let $\boldsymbol{u}$ be the 'unit word' $111 \cdots 1$. $u(x) = -g(x)g^{rev}(x)$ so $\boldsymbol{u} \in C$. But $x-1 \not| u(x)$ so $\boldsymbol{u} \notin D$. Similarly $-\boldsymbol{u} \notin D$, and $\boldsymbol{u}$, $-\boldsymbol{u}$ are in distinct cosets of D because their difference is $2\boldsymbol{u} = -\boldsymbol{u} \notin D$.
So applying Exercise 22, $C = D \cup (\boldsymbol{u}+D) \cup (-\boldsymbol{u}+D)$, and therefore any codeword $\mathbf{c}$ of C/D is of the form $\boldsymbol{a} \pm \boldsymbol{u}$ for some $\boldsymbol{a} \in D$.

Hence the weight of $\boldsymbol{c}$ is

$$
\begin{aligned}
\sum_{i=0}^{10} (a_i \pm 1)^2 & \qquad \text{from the proof of (ii)} \\
&= \sum a_i^2 \quad \pm\ 2\sum a_i \ +\ \sum 1 \\
&\equiv \text{weight of } \boldsymbol{a} \ \pm\ 0 \ +\ 11 \quad \text{from (i)} \\
&\equiv 0 \ \pm\ 0 \ +\ 11 \quad \text{from (ii)} \\
&\equiv 2
\end{aligned}
$$

$c(x)$ is a multiple of $g(x)$, say $c(x) = m(x)g(x)$. Then $c^{rev}(x) = m^{rev}(x)g^{rev}(x)$, so is a multiple of $g(x)g^{rev}(x)$. But $g(x)g^{rev}(x) = -u(x)$ so $c(x)c^{rev}(x) = 0,\ u(x)$ or $-u(x)$. $c^{rev}(1) = c(1) \neq 0$ since $\boldsymbol{c} \in C \backslash D$ so $x - 1 \not| c(x)$. So $c(x)c^{rev}(x) = \pm u(x)$, and therefore has weight 11. But $c(x)$ has weight w, so $c(x)c^{rev}(x)$ has at most w^2 non-zero coefficients, so $11 \leq w^2$. Hence $w \geq 4$, and since $w \equiv 2, w \geq 5$.

(iv) Now let $\boldsymbol{c}$ be any non-zero word of C, with weight w. From (ii) and (iii) $w \geq 3$ and if $w = 3$ then $\boldsymbol{c} \in D$. We now show that $w = 3$ is contradictory: suppose D had such a word $\boldsymbol{c}$ with $c(x) = \pm x^a \pm x^b \pm x^c (0 \leq a < b < c \leq 10)$. Then $c'(x) = x^{11-a}c(x) \equiv \pm 1 \pm x^i \pm x^j \in C$, where we have written i, j for $b-a$, $c-a$ respectively $(0 < i < j \leq 10)$. From (i) the sum of the three coefficients is $\equiv 0$ so they must all be $+1$ or all -1, so choose $c(x) = 1+x^i+x^j$. Now $c(x), c^{rev}(x)$ are multiples of $(x-1)g(x)$ and $g^{rev}(x)$ respectively, so $c(x)c^{rev}(x) \equiv x^{11} - 1 \equiv 0$.

$$\text{i.e.} \quad (1 + x^i + x^j)(x^{10} + x^{10-i} + x^{10-j}) \equiv 0$$

Expanding this and noting that $3x^{10} \equiv 0$, it reduces to

$$x^{10-i} + x^{10-j} + x^{10+i} + x^{10+j} + x^{10-i+j} + x^{10+i-j} \equiv 0.$$

These six terms must therefore consist of two cancelling triples. $10-i$ cannot be congruent to $10-j$ mod 11 since $0 < j-i < 11$, nor to $10+i$ as this implies $2i \equiv 0$ mod 11, nor to $10-i+j$ as this implies $j \equiv 0$ mod 11. Hence the only remaining candidate is $10-i \equiv 10+j \equiv 10+i-j$.

These imply $10 - i = 10 + j - 11$ and $10 + j = 11 + 10 + i - j$, from which $i + j = 11$ and $2_j - i = 11$, so $3j = 22$, which is impossible.

(v) From (ii) and (iv), if $\boldsymbol{c} \in D$ $w \geq 6$. From (iii), if $\boldsymbol{c} \notin D$ $w \geq 5$. Hence for all non-zero $\boldsymbol{c} \in C$ $w \geq 5$. But $g(x)$ has 5 non-zero terms, so represents a word of C with weight 5. Hence $d(C) = 5$

□

So $\langle g \rangle$ is a ternary cyclic [11, 6, 5] code, and if you now check the Hamming bound you will see that it is perfect, and therefore equivalent to the ternary 2-error correcting Golay code.

8.9 Exercises for Chapter 8

1. Work out $(2x^3 + x + 1)(x^4 + x^2 + 2x + 2)$ in $Z_3[X]/x^3 + 2x^2 + 1$.
2. Prove that $Z_p[X]/f(x)$ is a ring and that all its ideals are principal ideals.
3. Let $R = Z_2[X]/x^3 - 1$. Show that $\langle 1 + x \rangle = \langle 1 + x^2 \rangle$.
4. Show that the rows of the matrix G whose form is specified in Theorem 8.2 are independent
5. Which of the following codes are (a) cyclic, (b) linearly equivalent to a cyclic code?

   (i) {0000, 1100, 0110, 0011, 1001} over Z_2
   (ii) {0000, 1122, 2211} over Z_3
   (iii) the *q-ary* repetition code over Z_p, length n
   (iv) the set of all binary words of length n with even weight
   (v) the ternary code of length n whose codeword weights are all $\equiv 0$ mod 3
   (vi) as for (v) but with all codewords $c_0c_1c_2\cdots c_{n-1}$ satisfying $\sum c_i \equiv 0$ mod 3.

6.(a) $x^n - 1 = (x - 1)q(x)$ in $Z_2[X]$. What is $q(x)$?
   (b) Let g be the generator of the cyclic binary code C of length n. Show that if $x - 1|g(x)$ then all codewords have even weight.
   (c) Show that $q(x)$ (in (a)) is not a multiple of $x - 1$ if n is odd.
   (d) Let n (in (b)) be odd, and suppose C has a word of odd weight. Show that $111\cdots 1 \in C$ and that the set of all even weight words of C is a cyclic code having $(x - 1)g(x)$ as its generator.

7. Is $x^3 + 2x^2 + 2$ the generator of a cyclic ternary code of length 8 over Z_3?
8. Find all the binary cyclic codes of length 21 having dimension 9
9. What is the generator polynomial and dimension of the smallest ternary code containing 112110? What is its minimum distance?
10. Use the method of this section to find a nearly standard G for the cyclic binary [7, 4] code generated by $1 + x^2 + x^3$.
11. Given that $x^6 + x^5 + x^4 + x^3 + 1|x^{15} - 1$ over Z_2, find the codeword $\boldsymbol{m}G$ by the result of Theorem 8.8 where $\boldsymbol{m} = 010010001$.
12. With the code of Exercise 11 find the syndrome of the received word 010011000111010.
13. Following on from Exercises 11 and 12, find the syndromes of all the cyclic shifts of the word 010011000111010.

14. Verify that over Z_3 $g(x) = x^5 + x^4 + 2x^3 + x^2 + 2$ is a divisor of $x^{11} - 1$. Let C be the cyclic ternary [11, 6] code $\langle g(x) \rangle$.

    Given that $d(C) = 5$, use error trapping to decode the received word 20121020112. What proportion of errors of weight 2 are not correctable by this method?

15. Write $x^6 p(x^{-1})$ as a polynomial where

$$p(x) = 2x^6 + 3x^5 + x^3 + 4.$$

16. Find the generator polynomial for the cyclic member of Ham(3, 2) whose p.c. matrix is given in section 6.7.

17. Show that the cyclic code C is self orthogonal if and only if $\overline{h}(x)|g(x)$. Hence find a self orthogonal binary cyclic code with length 15.

18. Find a binary [15, 10] cyclic code which is all single, all double adjacent error correcting.

19. Let g be a $k \times n$ matrix whose first row is $g_0 g_1 \cdots g_{n-k} 00 \cdots 0$ where $g(x) = g_0 + g_1 x + \cdots + g_{n-k} x^{n-k}$ is a polynomial of degree $n - k$ with $g_0 \neq 0$. The remaining $k - 1$ rows are the first $k - 1$ cyclic shifts of this row. Let G' be the matrix constructed in the same way from g^{rev}. Show that G and G' generate equivalent codes.

20. There is a difficulty with this argument : there is nothing to prevent $j = k$, but in this case our cancelling pairs $ij, kl; kl, jl$ are not distinct, so the conclusion that $w^2 - w$ has to be a multiple of 4 is invalidated. Find a way out of this.

21. Using the (unproved) uniqueness remarks in Chapter 6 concerning the Golay codes, prove the corollary to Theorem 8.14.

22. Show that if C, D are linear codes over Z_3 with $D \subseteq C$ and $\dim(C) = 1 + \dim(D)$, then C is the union of three cosets of D.

9

The Reed–Muller family of codes

9.1 New codes from old

Many technical innovations are the result of combining desirable features of two or more gadgets to produce a composite object with even more desirable features. A natural question in coding theory is whether two good codes can be combined in some way to produce a better one. One such combination was invented by Plotkin and published in 1960. It can be used to describe the Reed–Muller codes, one of which was used in the NASA space explorations from 1969 to 1976, in particular to transmit the Mariner 9 pictures of Mars in January 1972.

9.2 Plotkin's construction

The method to be described takes two binary codes C_1 and C_2 of the same length n, and produces from them another binary code $C_1{*}C_2$ of length $2n$.

A typical word of $C_1{*}C_2$ is defined as follows: for the first n places take a word $\boldsymbol{u}$ of C_1, and for the last n take the word $\boldsymbol{u}+\boldsymbol{v}$ where $\boldsymbol{v}$ is any word of C_2. The complete code $C_1{*}C_2$ is the set of all words which can be formed in this way. If we use $\boldsymbol{a}|\boldsymbol{b}$ to represent the word formed by writing the bits of $\boldsymbol{b}$ after the bits of $\boldsymbol{a}$, then $C_1{*}C_2$ is given by:

Definition 9.1 $C_1{*}C_2 = \{\boldsymbol{u}|\boldsymbol{u}+\boldsymbol{v} \ : \ \boldsymbol{u} \in C_1, \boldsymbol{v} \in C_2\}$

For example, if $101011 \in C_1$ and $100010 \in C_2$, then the corresponding word of $C_1{*}C_2$ is

$$\begin{aligned} &1,0,1,0,1,1,1+1,0+0,1+0,0+0,1+1,1+0 \\ &= \ 101011001001. \end{aligned}$$

Notice that in general $C_1{*}C_2 \neq C_2{*}C_1$.

The following theorem relates some properties of C_1 and C_2 to those of $C_1{*}C_2$.

Theorem 9.1 If C_1 is binary linear $[n, k_1, d_1]$ and C_2 is binary linear $[n, k_2, d_2]$ then

(i) $C_1 * C_2$ is linear,

(ii) $\dim(C_1 * C_2) = k_1 + k_2$,

(iii) $d(C_1 * C_2) = \min\{2d_1, d_2\}$

Proof.

(i) This is left as an exercise [Ex 1]

(ii) Each ordered pair $(\boldsymbol{u}, \boldsymbol{v})$ of $C_1 \times C_2$ determines a word of $C_1 * C_2$ namely $\boldsymbol{u}|\boldsymbol{u} + \boldsymbol{v}$. Now $|C_1| = 2^{k_1}$ and $|C_2| = 2^{k_2}$ so $|C_1 \times C_2| = 2^{k_1} \times 2^{k_2} = 2^{k_1+k_2}$. If *these* $2^{k_1+k_2}$ *ordered pairs determine distinct words of* $C_1 * C_2$, this means $C_1 * C_2$ has $2^{k_1+k_2}$ codewords. Since $C_1 * C_2$ is linear from (i), its dimension is therefore $k_1 + k_2$.

So all that remains to complete the proof of (ii) is establish the underlined claim above. This is also left as an exercise. [Ex 2]

(iii) Use u_i, u_i', v_i, v_i' to represent the ith bits of $\boldsymbol{u}, \boldsymbol{u}', \boldsymbol{v}, \boldsymbol{v}'$ respectively.

Let $\boldsymbol{a} = \boldsymbol{u}|\boldsymbol{u} + \boldsymbol{v}$ and $\boldsymbol{b} = \boldsymbol{u}'|\boldsymbol{u}' + \boldsymbol{v}'$ be any two *different* words of $C_1 * C_2$.

There are two cases to consider: first, if $\boldsymbol{v} = \boldsymbol{v}'$ (so that $\boldsymbol{u} \neq \boldsymbol{u}'$ from the proof of (ii)), then $\boldsymbol{a} = u_1, u_2, \ldots, u_n, u_1 + v_1, u_2 + v_2, \ldots, u_n + v_n$ and $\boldsymbol{b} = u_1', u_2', \ldots, u_n', u_1' + v_1, u_2' + v_2, \ldots, u_n' + v_n$.

Now $\boldsymbol{u} + \boldsymbol{v}$ differs from $\boldsymbol{u}' + \boldsymbol{v}$ in exactly the same places as $\boldsymbol{u}$ differs from $\boldsymbol{u}'$, so $d(\boldsymbol{a}, \boldsymbol{b}) = 2d(\boldsymbol{u}, \boldsymbol{u}')$.

But $d(\boldsymbol{u}, \boldsymbol{u}') \geq d(C_1) = d_1$ so

$$d(\boldsymbol{a}, \boldsymbol{b}) \geq 2d_1 \qquad (1)$$

secondly, if $\boldsymbol{v} \neq \boldsymbol{v}'$ we have

$$\begin{aligned} d(\boldsymbol{a}, \boldsymbol{b}) &= d(\boldsymbol{u}, \boldsymbol{u}') + d(\boldsymbol{u} + \boldsymbol{v}, \boldsymbol{u}' + \boldsymbol{v}') \\ &= w(\boldsymbol{u} + \boldsymbol{u}') + w((\boldsymbol{u} + \boldsymbol{v}), (\boldsymbol{u}' + \boldsymbol{v}')) \\ &= d(\boldsymbol{o}, \boldsymbol{u} + \boldsymbol{u}') + w((\boldsymbol{u} + \boldsymbol{u}'), (\boldsymbol{v}' + \boldsymbol{v}')) \\ &= d(\boldsymbol{o}, \boldsymbol{u} + \boldsymbol{u}') + d(\boldsymbol{u} + \boldsymbol{u}', \boldsymbol{v} + \boldsymbol{v}') \\ &\geq d(\boldsymbol{o}, \boldsymbol{v} + \boldsymbol{v}') \text{ by the triangle inequality} \\ &\geq d(C_2) \\ &= d_2 \qquad (2) \end{aligned}$$

From results (1) and (2) above we have:

$$\begin{aligned} d(\boldsymbol{a}, \boldsymbol{b}) &\geq \text{ either } 2d_1 \text{ or } d_2 \\ \text{so } d(\boldsymbol{a}, \boldsymbol{b}) &\geq \min\{2d_1, d_2\} \\ \text{Hence } d(C_1 * C_2) &\geq \min\{2d_1, d_2\} \qquad (3) \end{aligned}$$

So all that remains to complete the proof of (iii) is to show that the

$\geq$ in result (3) above is in fact always $=$. We do this by finding two words of $C_1{*}C_2$, which differ in precisely $\min\{2d_1, d_2\}$ places, and again it is convenient to consider two separate cases.

First, suppose $2d_1 \geq d_2$ so that $\min\{2d_1, d_2\} = d_2$.

Let $\boldsymbol{a} = \boldsymbol{u}|\boldsymbol{u}+\boldsymbol{v}$ and $\boldsymbol{b} = \boldsymbol{u}|\boldsymbol{u}+\boldsymbol{v}'$ where $\boldsymbol{v}, \boldsymbol{v}'$ are chosen from C_2 to be at C_2's minimum separation d_2.

Then clearly $d(\boldsymbol{a}, \boldsymbol{b}) = d_2 = \min\{2d_1, d_2\}$.

Alternatively, if $2d_1 < d_2$ then $\min\{2d_1, d_2\} = 2d_1$.

In this case choose two words $\boldsymbol{u}, \boldsymbol{u}'$ of C_1, at C_1's minimum separation d_1, and let $\boldsymbol{a} = \boldsymbol{u}|\boldsymbol{u}+\boldsymbol{v}$ and $\boldsymbol{b} = \boldsymbol{u}'|\boldsymbol{u}'+\boldsymbol{u}$ where $\boldsymbol{v}$ is any word of C_2. Then clearly $\boldsymbol{u}+\boldsymbol{v}$ and $\boldsymbol{u}'+\boldsymbol{v}$ differ in exactly the same places as those where $\boldsymbol{u}$ and $\boldsymbol{u}'$ differ, so

$$d(\boldsymbol{a}, \boldsymbol{b}) = 2d_1$$

Hence, $d(C_1{*}C_2) = 2d_1 = \min\{2d_1, d_2\}$ □

[Ex 3, 4]

9.3 The Reed–Muller family

This two-parameter family of linear binary codes is conveniently described using the * construction. For each pair of parameters (r, m), the code $RM(r, m)$ has length 2^m and dimension r, and these parameters are constrained by $0 \leq r \leq m$. Because of this constraint, if we represent $RM(r, m)$ by a point (r, m) in the plane, those points which represent Reed–Muller codes are as shown in Figure 9.1.

Our first form of definition of the Reed–Muller family describes $RM(o, m)$ and $RM(m, m)$ explicitly, then gives any other $RM(r, m)$ (i.e. $0 < r < m$) in terms of $RM(r, m-1)$ and $RM(r-1, m-1)$.

To relate this to Figure 9.1: the circled codes are given explicitly, and to obtain, for example, $RM(3, 6)$, we need to know $RM(3, 5)$ and $RM(2, 5)$. Or in general, provided we know all the codes in one row, then all the codes in the row above can be determined. But the codes in the first two rows are known $(RM(0, 0), RM(0, 1)$ and $RM(1, 1))$. Hence the codes in the next row up can be found, then the next, ... and so on. This method of successively constructing each code from earlier codes in a list is very reminiscent of the way proof by induction works, and such constructions are called inductive (or recursive) definitions.

Here then is the definition:

Definition 9.2

1. $RM(0, m) = \{\mathbf{0}, \mathbf{1}\}$ where $\mathbf{0}$ is the all-zeros word and $\mathbf{1}$ is the all-ones word, of length 2^m.
2. $RM(m, m) = Z_2^{2^m}$.

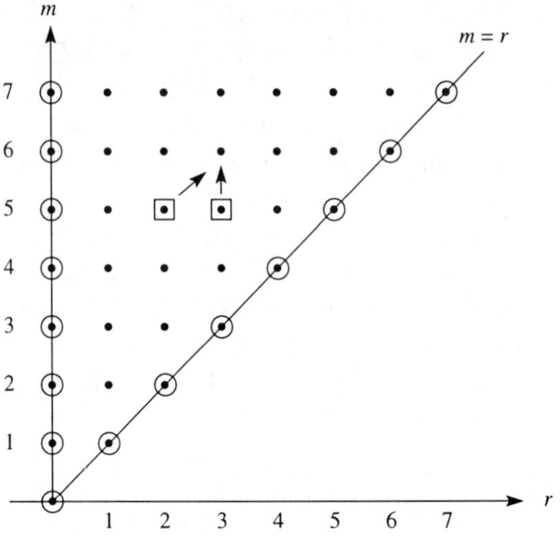

Figure 9.1 *The Reed–Muller codes.*

3. For $0 < r < m, RM(r,m) = RM(r,m-1) * RM(r-1,m-1)$.

So $RM(0,m)$ is simply the rather boring repetition code of length 2^m, with only two codewords. $RM(m,m)$ is the very big but useless code consisting of all binary strings of length 2^m, so this has 2^{2^m} words. Somewhere between these extremes we shall find some codes with good error correcting potential.

Let us first use this definition to see what the first few Reed–Muller codes look like.

From 2. (or 1.) $RM(0,0) = Z_2^1 = \{0,1\}$
From 1. $RM(0,1) = \{00,11\}$
From 2. $RM(1,1) = \{00,01,10,11\}$
From 1. $RM(0,2) = \{0000,1111\}$
From 3.
$$\begin{aligned} RM(1,2) &= RM(1,1) * RM(0,1) \\ &= \{00,01,10,11\} * \{00,11\} \\ &= \{00|00+00, 00|00+11, 01|01+00, 01|01+11, \\ &\quad 10|10+00, 10|10+11, 11|11+00, 11|11+11\} \end{aligned}$$
$\{0000, 0011, 0101, 0110, 1010, 1001, 1111, 1100\}$

Clearly this is an extremely messy process to continue much further, so you should now look at Exercises 5–7 which ask you to do one more step in the above construction, establish that for all r, m, $RM(r,m)$ is linear and has length 2^m, and find a generator matrix for $RM(r,m)$. [Ex 5–7]

The size of the Reed–Muller codes is an interesting number:

Theorem 9.2 $RM(r,m)$ has $2^{f(r,m)}$ codewords (that is, its dimension is $f(r,m)$), where $f(r,m) = \sum_{i=0}^{r} \binom{m}{i}$.

Proof. Since the RM family is defined inductively it should be no surprise that the proof is inductive. We use induction on m and take as our inductive hypothesis that $RM(r,m)$ has dimension $f(r,m)$ when $m = k$ for all $r \leq k$.

Basis step : IH certainly holds when $k = 0$, because the only Reed–Muller code with $m = 0$ is $RM(0,0)$ and its dimension is clearly 1, which agrees with

$$f(0,0) = \sum_{i=0}^{0} \binom{0}{i} = \binom{0}{0} = 1$$

Induction step : assuming IH, we try to prove that $RM(r, k+1)$ has dimension $f(r, k+1)$ for all $r \leq k+1$. There are three cases to consider, corresponding to the three clauses in the definition of $RM(r,m)$.

Case 1 : $r = k+1$. From clause 2, $RM(k+1, k+1)$ is the whole of $Z_2^{2^{k+1}}$, which has dimension 2^{k+1}. So we check that this agrees with $f(k+1, k+1)$.

$$f(k+1,k+1) = \sum_{i=0}^{k+1} \binom{k+1}{i} = \sum_{i=0}^{k+1} \binom{k+1}{i} 1^{k+1-i} 1^{i} \text{ and}$$

this last expression is just the binomial expansion of $(1+1)^{k+1}$, that is 2^{k+1} as required.

Case 2: $r = 0$. From clause 1, $RM(0, k+1)$ is just the repetition code of length 2^{k+1}, with dimension 1. This agrees with $f(0, k+1) = \sum_{i=0}^{0} \binom{k+1}{i}$ $= \binom{k+1}{0} = 1$.

Case 3 : $0 < r < k+1$. In this case clause 3 applies, and we have $RM(r, k+1) = RM(r,k) * RM(r-1,k)$.

By IH we have

$$\begin{aligned} \dim(RM(r,k)) &= f(r,k) \\ \text{and } \dim(RM(r-1,k)) &= f(r-1,k) \end{aligned}$$

So by Theorem 9.1 part(ii) we have

$$\dim(RM(r,k+1)) = f(r,k) + f(r-1,k).$$

So it remains to show that this is equal to $f(r, k+1)$, and this is just a matter of manipulating binomial coefficients:

$$\begin{aligned} f(r,k) + f(r-1,k) &= \binom{k}{0} + \binom{k}{1} + \binom{k}{2} + \ldots + \binom{k}{r-1} + \binom{k}{r} \\ &\quad + \binom{k}{0} + \binom{k}{1} + \ldots + \binom{k}{r-2} + \binom{k}{r-1} \\ &= \binom{k}{0} + \binom{k+1}{1} + \binom{k+1}{2} + \ldots + \binom{k+1}{r-1} + \binom{k+1}{r} \quad (*) \\ &= \binom{k+1}{0} + \binom{k+1}{1} + \binom{k+1}{2} + \ldots + \binom{k+1}{r-1} + \binom{k+1}{r} \quad (**) \\ &= \sum_{i=0}^{r} \binom{k+1}{i} = f(r,k+1) \text{ as required} \end{aligned}$$

Note: line (*) comes from adding the pairs of terms vertically aligned in the previous two lines and using the Pascal triangle relation. Line(**) simply uses the fact that $\binom{n}{0}$ is independent of n to replace $\binom{k}{0}$ by $\binom{k+1}{0}$ and hence prepare for the final line.

□

Theorem 9.3 The minimum distance of $RM(r, m)$ is 2^{m-r}.

The proof is left as an exercise. □

[Ex 8, 9]

From the generator matrices established in Exercise 7 we can deduce a nice connection between the codes in a given row of Figure 9.1. It is simply that any $RM(r, m)$ code is a subcode of any code to the right of it in the same row. Specifically, we prove

Theorem 9.4 For all m, and all positive $r \leq m$, $RM(r - 1, m) \subset RM(r, m)$.

Proof. Again the proof is by induction on m. We take as the I H the assertion that the theorem is true for $m \leq k$. For $m = 1$ we just have to show that $RM(0, 1) \subset RM(1, 1)$, and this is clearly true by referring to our listing of the first few Reed–Muller codes. That completes the basis of the induction, so now we have to take any word in $RM(r - 1, k + 1)$ and show that it is in $RM(r, k + 1)$.

Now

$$\boldsymbol{w} \in RM(r - 1, k + 1) \Rightarrow \boldsymbol{w} = \boldsymbol{u}|\boldsymbol{u} + \boldsymbol{v}$$

where

$$\boldsymbol{u} \in RM(r - 1, k), \boldsymbol{v} \in RM(r - 2, k).$$

But by I H $\boldsymbol{u} \in RM(r, k)$ and $\boldsymbol{v} \in RM(r-1, k)$. Hence $\boldsymbol{w} \in RM(r, k+1)$ by using the definition of the Reed–Muller codes again. □

[Ex 10]

9.4 An alternative description of Reed–Muller codes

This makes use of some elementary Boolean algebra. We shall see in this section how the code words of a Reed–Muller code may be identified with a set of Boolean polynomials. What we need is a modification of the disjunctive normal form which uses the 'exclusive or', $\oplus$, rather than the 'inclusive or', $+$. The relations between these, and the properties of $\oplus$ which we require are listed below:

1	$a \oplus (b \oplus c)$	$=$	$(a \oplus b) \oplus c$
2	$a + b$	$=$	$a \oplus b \oplus ab$
3	$a \oplus b$	$=$	$b \oplus a$
4	$1 \oplus a$	$=$	$\bar{a}$
5	$0 \oplus a$	$=$	a
6	$a \oplus a$	$=$	0
7	$a(b \oplus c)$	$=$	$ab \oplus ac$

x	y	z	$f(x,y,z)$
0	0	0	1
1	0	0	0
0	1	0	0
1	1	0	0
0	0	1	0
1	0	1	1
0	1	1	0
1	1	1	0

Figure 9.2 *Obtaining a* $\oplus$ *normal form.*

If we have a function f, of m Boolean variables, its truth table will have 2^m rows. The last column, giving the values of $f(x_1, \ldots, x_m)$, is then a binary string of length 2^m, and there are $2^{(2^m)}$ such strings, corresponding to the fact that there are $2^{(2^m)}$ Boolean functions of m variables.

The following example shows a way of starting from a Boolean function given by its truth table and constructing an expression for it as a $\oplus$ sum of products of the variables. Let f be the Boolean function given by the truth table of Figure 9.2.

The disjunctive normal form of $f(x, y, z)$ is therefore $\bar{x}\bar{y}\bar{z} + x\bar{y}z$. Writing this in terms of $\oplus$ using the seven properties listed above we get:

$$\begin{aligned} & (1 \oplus x)(1 \oplus y)(1 \oplus z) + x(1 \oplus y)z \\ = \; & (1 \oplus x)(1 \oplus y)(1 \oplus z) \oplus x(1 \oplus y)z \\ = \; & 1 \oplus x \oplus y \oplus z \oplus xy \oplus xz \oplus yz \oplus xyz \oplus xz \oplus xyz \\ = \; & 1 \oplus x \oplus y \oplus z \oplus xy \qquad\quad \oplus yz \end{aligned}$$

[Ex 11, 12]

There are a couple of things which make this $\oplus$ form particularly convenient. One is that $\oplus$ is an easy Boolean operation to implement with electronic hardware (though this book does not consider such problems). The other is that it is unique, by which we mean that any $\oplus$ sum equivalent to $\bar{x}\bar{y}\bar{z} + x\bar{y}z$ must contain as its individual terms $1, x, y, z, xy, yz$ and no others. The only freedom we have is in the order in which the terms are written down. (This is reminiscent of the unique prime factorization theorem of Chapter 3.) Note that we are using *term* to mean any Boolean product of any subset of the variables, conventionally letting the constant term 1 correspond to choosing the empty subset. We do not count the other constant 0 as a term since this would invalidate the uniqueness claim we made above, since $x \oplus xy \oplus z = x \oplus xy \oplus z \oplus 0$.

We define the *degree* of a term to be the number of variables in it, so $1, x_1, x_4, x_5x_3$ have degrees $0, 1, 1, 2$ respectively, and it should be clear from the example following Figure 9.2 or from Exercise 12 that any Boolean function of m variables can be expressed as a $\oplus$ sum of terms each of degree at most m. It remains to establish the uniqueness claim, and this is surprisingly easy, just a counting argument.

Theorem 9.5 Each Boolean function of m variables has a unique expression as a $\oplus$ sum of terms of degree $\leq m$.

Proof. Each subset of the m variables corresponds to a term. Hence there are 2^m possible terms. There are $2^{(2^m)}$ Boolean functions of m variables and there are $2^{(2^m)}$ subsets of the 2^m terms. For each of these subsets consider the function obtained from the $\oplus$ sum of its members (with the constant zero function corresponding to choosing the empty set). So the number of $\oplus$ sums is exactly the same as the number of Boolean functions, so no function can be represented by more than one $\oplus$ sum, and this is precisely our uniqueness claim. □

Notice that there is no such neat uniqueness result with $\oplus$ replaced by $+$, since, for example,

$$xy + xz + y = y + xz.$$

To relate this discussion to binary codes, note that $\oplus$ is just addition modulo 2 since $0 \oplus 0 = 1 \oplus 1 = 0$ and $0 \oplus 1 = 1 \oplus 0 = 1$. Furthermore, as a binary vector, a term corresponds to the binary column of its truth table, so in the case of functions of 3 variables the constant term 1 is $(11111111)^T, xz$ is $(00000101)^T$, etc.

We can therefore think of the individual terms (from a set of m variables) as an independent set which spans the whole of $Z_2^{2^m}$, so they make a basis for this space.

Our purpose now is to show that $RM(r, m)$ is, in this interpretation, just the set of all $\oplus$ sums of Boolean terms of degree at most r, over the variables $x_1, x_2, \ldots, x_m$. We first check that we do have the right numbers of functions and terms. Clearly there are $\binom{m}{i}$ terms of degree i, so the number of terms of degree $\leq r$ is $\sum_{i=0}^{r}\binom{m}{i}$. This, from Theorem 9.2, is precisely the dimension of $RM(r, m)$, so it looks as if we are on the right lines.

To make the proof run smoothly we stick to the ordering of the rows and columns of truth tables used thus far. That is, the column for x_1 is $[010101\ldots01]^T, x_2$ has $[001100110\ldots, 0011]^T$, until x_m which just has a single block of 2^{m-1} zeros followed by 2^{m-1} 1s. If we did not have this ordering the proof would be harder to follow, and we might end up with a code equivalent but not identical to what we have previously defined as $RM(r, m)$.

The proof works essentially by showing that the set $BF(r, m)$ of Boolean functions of m variables whose $\oplus$ sum forms contain only terms of degree at most r satisfy exactly the same recurrence relations as do the set of codewords of $RM(r, m)$.

We first need a preliminary result about Boolean functions.

Theorem 9.6 If f is a Boolean function of m variables $x_1, x_2, \ldots, x_m$, then there exist Boolean functions g, h of $m-1$ variables such that

$$f(x_1, x_2, \ldots, x_m) = x_m.g(x_1, x_2, \ldots, x_{m-1}) \oplus h(x_1, x_2, \ldots, x_{m-1})$$

Proof.

$$\begin{aligned} f(x_1, \ldots, x_m) &= x_m f(x_1, \ldots, x_{m-1}, 1) &+& \quad \bar{x}_m.f(x_1, \ldots, x_{m-1}, 0) \\ &= x_m f(x_1, \ldots, x_{m-1}, 1) &\oplus& \quad \bar{x}_m.f(x_1, \ldots, x_{m-1}, 0) \\ &= x_m f(x_1, \ldots, x_{m-1}, 1) &\oplus& \quad (1 \oplus x_m).f(x_1, \ldots, x_{m-1}, 0) \\ &= x_m f(x_1, \ldots, x_{m-1}, 1) &\oplus& \quad x_m.f(x_1, \ldots, x_{m-1}, 0) \\ & &\oplus& \quad f(x_1, \ldots, x_{m-1}, 0) \\ &= x_m[f(x_1, \ldots, x_{m-1}, 1) &\oplus& \quad f(x_1, \ldots, x_{m-1}, 0)] \\ & &\oplus& \quad f(x_1, \ldots, x_{m-1}, 0) \\ &= x_m.g(x_1, \ldots, x_{m-1}) &\oplus& \quad h(x_1, \ldots, x_{m-1}) \end{aligned}$$

where g and h are defined by

$$\begin{aligned} g(x_1, \ldots, x_{m-1}) &= f(x_1, \ldots, x_{m-1}, 1) \quad \oplus \quad f(x_1, \ldots, x_{m-1}, 0) \\ \text{and } h(x_1, \ldots, x_{m-1}) &= f(x_1, \ldots, x_{m-1}, 0) \end{aligned}$$

□

Now suppose that $f(x_1, \ldots, x_m)$ has degree $\leq r$. It could happen that $f(x_1, \ldots, x_m)$ has terms of degree r in its $\oplus$ expansion and that one or more of these terms does not involve x_m. So all that we can guarantee about the degree of $h(x_1, \ldots, x_{m-1}) = f(x_1, \ldots, x_{m-1}, 0)$ is that it too is $\leq r$. So, as a binary vector $h(x_1, \ldots, x_{m-1}) \in BF(r, m-1)$. We shall call this the vector $\boldsymbol{h}$ (of length 2^{m-1}).

Now consider

$$x_m g(x_1, \ldots, x_{m-1}) = x_m[f(x_1, \ldots, x_{m-1}, 0) \oplus f(x_1, \ldots, x_{m-1}, 1)].$$

This is part of the $\oplus$ expansion of $f(x_1, \ldots, x_m)$ from Theorem 9.6, so it has degree $\leq r$, and therefore $g(x_1, \ldots, x_{m-1})$ can only have degree $\leq r-1$. The corresponding binary vector $\boldsymbol{g}$ is therefore a member of $BF(r-1, m-1)$.

All this is illustrated in the truth table of Figure 9.3. We take $f(x_1, \ldots, x_m)$ to correspond to the vector $\boldsymbol{f} = a_1 a_2 \ldots a_{2^{m-1}} a'_1 a'_2 \ldots a'_{2^{m-1}}$, and the entries in the three columns to the right of this are then deduced from the defining formulae at the top of these columns.

From the result of Theorem 9.6 and Figure 9.3 we see that $\boldsymbol{f} = \boldsymbol{o}|\boldsymbol{g} \oplus \boldsymbol{h}|\boldsymbol{h} = \boldsymbol{h}|\boldsymbol{h} \oplus \boldsymbol{g}$.

Now $\boldsymbol{f}$ was an arbitrary member of $BF(r, m)$, and $\boldsymbol{h}$, $\boldsymbol{g}$ are members of $BF(r, m-1)$, $BF(r-1, m-1)$ respectively. So any word in $BF(r, m)$ is a member of $BF(r, m-1) * BF(r-1, m-1)$, as defined in section 9.2. Conversely, it is clear that if we take arbitrary $\boldsymbol{h}$, $\boldsymbol{g}$ in $BF(r, m-1)$, $BF(r-1, m-1)$ respectively, and form the word $\boldsymbol{h}|\boldsymbol{h} \oplus \boldsymbol{g}$, the result is in $BF(r, m)$. Hence $BF(r, m) = BF(r, m-1) * BF(r-1, m-1)$.

When $r = 0$ $BF(r-1, m-1)$ does not exist so this formula cannot be used. However $BF(o, m)$ consists of just the constant functions of m

$$f(x_1,\ldots,x_m) =$$
$$x_m(f(x_1,\ldots,x_{m-1},1) \oplus f(x_1,\ldots,x_{m-1},0)) \oplus f(x_1,\ldots,x_{m-1},0)$$

x_1	x_2	$\ldots$	x_{m-1}	x_m	I	II	III	IV
0	0		0	0	a_1	0	a_1	$a_1' \oplus a_1$
1	0		0	0	a_2	0	a_2	$a_2' \oplus a_2$
0	1		$\vdots$	0	$\vdots$	0	$\vdots$	$\vdots$
1	1		$\vdots$	$\vdots$	$\vdots$	$\vdots$	$\vdots$	$\vdots$
$\vdots$	$\vdots$		0	$\vdots$	$\vdots$	$\vdots$	$\vdots$	$\vdots$
$\vdots$	$\vdots$		1	$\vdots$	$\vdots$	$\vdots$	$\vdots$	$\vdots$
0	0		1	$\vdots$	$\vdots$	$\vdots$	$\vdots$	$\vdots$
1	0		$\vdots$	$\vdots$	$\vdots$	$\vdots$	$\vdots$	$\vdots$
0	1		$\vdots$	$\vdots$	$\vdots$	$\vdots$	$\vdots$	$\vdots$
1	1		1	0	$a_{2^{m-1}}$	0	$a_{2^{m-1}}$	$a'_{2^{m-1}} \oplus a_{2^{m-1}}$
0	0		0	1	a_1'	$a_1' \oplus a_1$	a_1	$a_1' \oplus a_1$
1	0		0	1	a_2'	$a_2' \oplus a_2$	a_2	$a_2' \oplus a_2$
0	1		$\vdots$	1	$\vdots$	$\vdots$	$\vdots$	$\vdots$
1	1		$\vdots$	$\vdots$	$\vdots$	$\vdots$	$\vdots$	$\vdots$
$\vdots$	$\vdots$		0	$\vdots$	$\vdots$	$\vdots$	$\vdots$	$\vdots$
$\vdots$	$\vdots$		1	$\vdots$	$\vdots$	$\vdots$	$\vdots$	$\vdots$
0	0		1	$\vdots$	$\vdots$	$\vdots$	$\vdots$	$\vdots$
1	0		$\vdots$	$\vdots$	$\vdots$	$\vdots$	$\vdots$	$\vdots$
0	1		$\vdots$	$\vdots$	$\vdots$	$\vdots$	$\vdots$	$\vdots$
1	1		1	1	$a'_{2^{m-1}}$	$a'_{2^{m-1}} \oplus a_{2^{m-1}}$	$a_{2^{m-1}}$	$a'_{2^{m-1}} \oplus a_{2^{m-1}}$
					$\downarrow$	$\downarrow$	$\downarrow$	$\downarrow$
					$\boldsymbol{f}$	$\boldsymbol{o}\vert\boldsymbol{g}$	$\boldsymbol{h}\vert\boldsymbol{h}$	$\boldsymbol{g}\vert\boldsymbol{g}$

Column I = values of $f(x_1, x_2, \ldots, x_m)$
Column II = values of $x_m f(x_1, x_2, \ldots, x_{m-1}, 1)$
$\oplus x_m f(x_1, x_2, \ldots, x_{m-1}, 0)$
Column III = values of $f(x_1, x_2, \ldots, x_{m-1}, 0)$
Column IV = values of $f(x_1, x_2, \ldots, x_{m-1}, 1) \oplus f(x_1, x_2, \ldots, x_{m-1}, 0)$

Figure 9.3

variables, and as binary vectors these correspond to just the all zero and all one words of length 2^m. Note that this agrees with our previous definition of $RM(o, m)$. [Ex 13–15]

When defining the codes $RM(r, m)$ we imposed the restriction that $r \leq m$ and defined $RM(m, m)$ explicitly as the set of all words of $Z_2^{2^m}$. This was necessary because attempting to use the recurrence relation to determine $RM(m, m)$ would necessitate using $RM(m, m-1)$ which is not defined due to the restriction $r \leq m$. With $BF(r, m)$ no such restriction is necessary, and the formula with $r = m$, namely $BF(m, m) = BF(m, m-1) * BF(m-1, m-1)$, still holds, as you can check using the results of Exercises 13–15.

In any case we have now succeeded in showing that $RM(r, m)$ and $BF(r, m)$ coincide when $r = 0$ and when $r = m$, and for values of r between these extremes they satisfy the same recurrence relation. Hence they are identical for all r with $0 \leq r \leq m$, and we therefore have two equivalent descriptions of the Reed–Muller codes. [Ex 16]

A further advantage of our new description is that it is easy to find a generator matrix for any given $RM(r, m)$ without working through the smaller Reed–Muller codes and using the recursive formula developed in Exercise 7. The key to this is our previous remark that $BF(r, m)$ has as a basis the set of all terms of degree $\leq r$, with m variables available. To illustrate for $RM(2, 4)$, suppose a truth table for four variables is written in the standard way. Then a basis for $RM(2, 4)$ is the set of columns corresponding to $1, x_1, x_2, \ldots, x_4, x_1x_2, x_1x_3, \ldots, x_3x_4, x_1x_2x_3, \ldots, x_1x_2x_3x_4$. Writing these columns as rows of the generator matrix we obtain a generator matrix for $RM(2, 4)$ as:

1	1	1	1	1	1	1	1	1	1	1	1	1	1	1	1	1
x_1	0	1	0	1	0	1	0	1	0	1	0	1	0	1	0	1
x_2	0	0	1	1	0	0	1	1	0	0	1	1	0	0	1	1
x_3	0	0	0	0	1	1	1	1	0	0	0	0	1	1	1	1
x_4	0	0	0	0	0	0	0	0	1	1	1	1	1	1	1	1
x_1x_2	0	0	0	1	0	0	0	1	0	0	0	1	0	0	0	1
x_1x_3	0	0	0	0	0	1	0	1	0	0	0	0	0	1	0	1
x_1x_4	0	0	0	0	0	0	0	0	0	1	0	1	0	1	0	1
x_2x_3	0	0	0	0	0	0	1	1	0	0	0	0	0	0	1	1
x_2x_4	0	0	0	0	0	0	0	0	0	0	1	1	0	0	1	1
x_3x_4	0	0	0	0	0	0	0	0	0	0	0	0	1	1	1	1
$x_1x_2x_3$	0	0	0	0	0	0	0	1	0	0	0	0	0	0	0	1
$x_1x_2x_4$	0	0	0	0	0	0	0	0	0	0	0	1	0	0	0	1
$x_1x_3x_4$	0	0	0	0	0	0	0	0	0	0	0	0	0	1	0	1
$x_2x_3x_4$	0	0	0	0	0	0	0	0	0	0	0	0	0	0	1	1
$x_1x_2x_3x_4$	0	0	0	0	0	0	0	0	0	0	0	0	0	0	0	1

9.5 Hamming codes, first order Reed–Muller codes – some connections

If we restrict attention to first order Reed–Muller codes there is yet another way of obtaining them – as duals of extended Hamming codes.

To understand this relationship start with a parity check matrix H_m for a binary Hamming code of length $2^m - 1$. Recall that this is a $(2^m - 1) \times m$ matrix whose columns are the binary representations of the numbers from 1 to $2^m - 1$ in any order. Now consider this code extended by adding a single parity check bit at the end of each code word and recall from Chapter 6 that a p.c. matrix for the extended code can be obtained by adding a column of zeros to the right hand end of H_m, then adding a row of ones at the top of the result. If we call the matrices constructed in these two steps B_m and J_m the results for $m = 3$ are shown below.

$$H_3 = \begin{bmatrix} 0 & 0 & 0 & 1 & 1 & 1 & 1 \\ 0 & 1 & 1 & 0 & 0 & 1 & 1 \\ 1 & 0 & 1 & 0 & 1 & 0 & 1 \end{bmatrix}, \qquad B_3 = \begin{bmatrix} 0 & 0 & 0 & 1 & 1 & 1 & 1 & 0 \\ 0 & 1 & 1 & 0 & 0 & 1 & 1 & 0 \\ 1 & 0 & 1 & 0 & 1 & 0 & 1 & 0 \end{bmatrix},$$

$$J_3 = \begin{bmatrix} 1 & 1 & 1 & 1 & 1 & 1 & 1 & 1 \\ 0 & 0 & 0 & 1 & 1 & 1 & 1 & 0 \\ 0 & 1 & 1 & 0 & 0 & 1 & 1 & 0 \\ 1 & 0 & 1 & 0 & 1 & 0 & 1 & 0 \end{bmatrix},$$

or in general,

$$J_m = \left[\frac{\mathbf{1}}{H_m\mathbf{0}}\right] = \left[\frac{\mathbf{1}}{B_m}\right]$$

where $\mathbf{0}$ and $\mathbf{1}$ are the all zero column and all one row of lengths $m, 2^m$ respectively.

J_m is a generator matrix for the dual of an extended Hamming code, and we have the tools to prove:

Theorem 9.7 $RM(1, m)$ is the dual of an extended binary Hamming code of length 2^m.

Proof. Clearly from the preceding discussion we just need to show that J_m is a generator for a code equivalent to $RM(1, m)$. To do this we first need to remove the ambiguity in the definition of J_m, so just agree a fixed order for the columns of H_m. For definiteness, choose column i to be the binary representation of the integer i $(1 \leq i \leq 2^m - 1)$, though any agreed order would do.

Let $G(1, m)$ be the generator matrix constructed by Exercise 7. We show that the matrices J_m and $G(1, m)$ differ only by a permutation of their columns, so that the codes they generate are cetainly equivalent. The notation $A \sim B$ will be used for matrices which differ at most in this way.

Now J_1 is $\begin{bmatrix} 1 & 1 \\ 0 & 1 \end{bmatrix}$ and in our solution to Exercise 7 we took $G(1,1)$ to be $\begin{bmatrix} 1 & 1 \\ 0 & 1 \end{bmatrix}$, so certainly $J_1 \sim G(1,1)$.

For $m > 1$ we had

$$G(1,m) = \begin{bmatrix} G(1,m-1) & G(1,m-1) \\ \mathbf{0} & \mathbf{1} \end{bmatrix}$$

from which we see that the top row of $G(1,m)$ is $\mathbf{1}$ for all m.

Now

$$J_m = \begin{bmatrix} \mathbf{1} \\ B_m \end{bmatrix}$$

and exactly half the columns of B_m end in 0 so by doing a column permutation on J_m, put these columns in the first 2^{m-1} places. This of course makes no difference to the top row so the resulting matrix can be partitioned as:

$$J_m \sim \begin{bmatrix} 111\ldots1 & 111\ldots1 \\ M_1 & M_2 \\ 000\ldots0 & 111\ldots1 \end{bmatrix}$$

where $M_1 \sim M_2$. A further column permutation confined to the last 2^{m-1} columns will convert M_2 into a copy of M_1 so

$$J_m \sim \begin{bmatrix} 111\ldots1 & 111\ldots1 \\ M_1 & M_1 \\ 000\ldots0 & 111\ldots1 \end{bmatrix}$$

and finally, column permutations to bring the all zero column of each copy of M_1 to its right hand end will achieve

$$J_m \sim \begin{bmatrix} 111\ldots1 & & 111\ldots1 & \\ H_{m-1} & \mathbf{0} & H_{m-1} & \mathbf{0} \\ \mathbf{0} & & \mathbf{1} & \end{bmatrix} = \begin{bmatrix} J_{m-1} & J_{m-1} \\ \mathbf{0} & \mathbf{1} \end{bmatrix}$$

If we now take as the induction hypothesis the assertion that $J_{m-1} \sim G(1,m-1)$ we have

$$J_m \sim \begin{bmatrix} J_{m-1} & J_{m-1} \\ \mathbf{0} & \mathbf{1} \end{bmatrix} \sim \begin{bmatrix} G(1,m-1) & G(1,m-1) \\ \mathbf{0} & \mathbf{1} \end{bmatrix} = G(1,m)$$

and the proof is complete. □

Here is another way of obtaining the same result, using the parameters of the Reed–Muller codes derived in this chapter and the uniqueness results for Hamming codes from Chapter 6. We first show that the dual of any Reed–Muller code is another Reed–Muller code. Specifically, we have:

Theorem 9.8 $RM(r,m)^{\perp} = RM(m-r-1,m)$.

Proof. The first step is to show that every word in $RM(r,m)$ is orthogonal to each word of $RM(m-r-1,m)$. In Chapter 5 you saw that this is equivalent to showing that every row of any generator matrix for $RM(r,m)$ is orthogonal to each row of a generator for $RM(m-r-1,m)$. The proof of this is by induction on m.

Suppose the result holds for $m=k-1$ and all relevant r. $m=1$ is the smallest value of m for which both matrices are meaningful, and we leave you to check the basis step. [Ex 17]

Now using our recursively defined generator matrices we have:

$$G(r,k) = \begin{bmatrix} G(r,k-1) & G(r,k-1) \\ \mathbf{0} & G(r-1,k-1) \end{bmatrix} = \begin{bmatrix} A & A \\ 0 & B \end{bmatrix}$$

and

$$G(k-r-1,k) = \begin{bmatrix} G(k-r-1,k-1) & G(k-r-1,k-1) \\ \mathbf{0} & G(k-r-2,k-1) \end{bmatrix} = \begin{bmatrix} C & C \\ 0 & D \end{bmatrix}$$

Let us denote typical words from the AA, OB, CC, OD sections of the partitions by $\boldsymbol{a}|\boldsymbol{a}, \boldsymbol{o}|\boldsymbol{b}, \boldsymbol{c}|\boldsymbol{c}, \boldsymbol{o}|\boldsymbol{d}$ respectively.

We need to show that $(\boldsymbol{a}|\boldsymbol{a}).(\boldsymbol{c}|\boldsymbol{c}), (\boldsymbol{a}|\boldsymbol{a}).(\boldsymbol{o}|\boldsymbol{d}), (\boldsymbol{o}|\boldsymbol{b}).(\boldsymbol{c}|\boldsymbol{c}), (\boldsymbol{o}|\boldsymbol{b}).(\boldsymbol{o}|\boldsymbol{d})$ are all zero.

Now

$(\boldsymbol{a}|\boldsymbol{a}).(\boldsymbol{c}|\boldsymbol{c}), = \boldsymbol{a}.\boldsymbol{c} + \boldsymbol{a}.\boldsymbol{c} = 0$ (remember the arithmetic is mod 2).

$(\boldsymbol{a}|\boldsymbol{a}).(\boldsymbol{o}|\boldsymbol{d}) = \boldsymbol{a}.\boldsymbol{d} = 0$ by the induction hypothesis since $\boldsymbol{a} \in RM(r,k-1)$ and $\boldsymbol{d} \in RM(k-r-2,k-1)$

$(\boldsymbol{o}|\boldsymbol{b}).(\boldsymbol{c}|\boldsymbol{c}) = \boldsymbol{b}.\boldsymbol{c} = 0$ again by the induction hypothesis because $\boldsymbol{b} \in RM(r-1,k-1)$ and $\boldsymbol{c} \in RM(k-r-1,k-1)$

$(\boldsymbol{o}|\boldsymbol{b}).(\boldsymbol{o}|\boldsymbol{d}) = \boldsymbol{b}.\boldsymbol{d} = 0$ because $\boldsymbol{b} \in RM(r-1,k-1) \subset RM(r,k-1)$ and $\boldsymbol{d} \in RM(k-r-2,k-1)$ so the IH yields what we want again.

So we have now established that $RM(m-r-1,m) \subseteq RM(r,m)^{\perp}$. Finally, to change $\subseteq$ to $=$ we prove that these two codes have the same dimension, so by section 5.3, item 11. they must be equal. This is left as an exercise, with the hint that it is just another manipulation of binomial coefficients. [Ex 18]

□

Specializing to first order codes we have $RM(1,m)^{\perp} = RM(m-2,m)$, and we now compare $RM(m-2,m)$ with the extended binary Hamming code of length 2^m. This code, recall from Chapter 6, has dimension $2^m - 1 - m$ and minimum distance 4.

Now $RM(m-2,m)$ also has length 2^m; its dimension is:

$$\begin{array}{rcccccccl} \sum_{i=0}^{m-2}\binom{m}{i} & = & \sum_{i=0}^{m}\binom{m}{i} & - & \binom{m}{m} & - & \binom{m}{m-1} & & \\ & = & 2^m & - & 1 & - & m\,; & & \text{and} \\ d(RM(m-2,m)) & = & 2^{m-(m-2)} = 4, & & \text{by Theorem 9.3.} & & & & \end{array}$$

So $RM(1,m)^{\perp}$ is a code with the same parameters as the extended Hamming code of length 2^m, so by the uniqueness result of Chapter 6 the two codes are equivalent and we have our alternative proof of Theorem 9.7. □

9.6 Exercises for Chapter 9

1. Prove part (i) of Theorem 9.1.
2. Complete the proof of (ii) by showing that if $\boldsymbol{u}, \boldsymbol{u}' \in C_1$ and $\boldsymbol{v}, \boldsymbol{v}' \in C_2$ then the words $\boldsymbol{u}|\boldsymbol{u}+\boldsymbol{v}$ and $\boldsymbol{u}'|\boldsymbol{u}'+\boldsymbol{v}'$ of $C_1 * C_2$ cannot be equal, unless of course $\boldsymbol{u} = \boldsymbol{u}'$ and $\boldsymbol{v} = \boldsymbol{v}'$.
3. Let C_1 and C_2 be binary linear codes of length n with generator matrices G_1, G_2, respectively. Show that a generator matrix for $C_1 * C_2$ is
$$\begin{bmatrix} G_1 & G_1 \\ 0 & G_2 \end{bmatrix}$$
where 0 represents the all-zero matrix of appropriate size.
4. Let C be a code of the family Ham (3,2), and D = C * C.
   (a) Find the dimension and minimum distance of D.
   (b) Is D a perfect code?
   (c) Give two reasons why, whichever Slepian array is chosen for D, not all words of weight 2 can be coset leaders.
   (d) Find a parity check matrix for D, and hence find two error patterns each of weight 2, only one of which can be correctly decoded.
5. List the code words of $RM(2,3)$ using our initial definition of $RM(r,m)$.
6. Show that for all r, m, $RM(r,m)$ is linear and has length 2^m.
7. If $G(r,m)$ is a generator matrix for $RM(r,m)$ show that $G(m,m)$ may be taken to be
$$\begin{bmatrix} G(m-1,m) \\ 0000\ldots01 \end{bmatrix}$$
and $G(0,m) = [111\ldots1]$ (of length 2^m), and find a recursive definition of $G(r,m)$ for $0 < r < m$.

8. By copying the technique for proving Theorem 9.2 and making use of Theorem 9.1 (part (iii)), prove Theorem 9.3.
9. Using Exercise 7 construct generator matrices for $RM(2,3)$ and $RM(2,4)$.
10. Show that there are Reed–Muller codes which are at least 7-error correcting and contain over 1000 codewords. Find one of shortest length.
11. Check that you can follow the derivation of the $\oplus$ form of $x\bar{y}z + \bar{x}\bar{y}\bar{z}$ by saying which of the rules 1–7 is being used at each step.
12. Derive the $\oplus$ form of the Boolean function, $xyz + xy\bar{z} + x\bar{y}\bar{z} + \bar{x}\bar{y}z$.
13. If C is any binary linear code, find a simple description of the code C * C.
14. Show from its definition that $BF(r,m)$ is linear.
15. What is $BF(r,m)$ when $r \geq m$?
16. Show that Theorems 9.2 and 9.4 have particularly easy proofs if the Boolean function description of $RM(r,m)$ is used.
17. Carry out the basis step in the proof of Theorem 9.8.
18. Show that $\dim[RM(m-r-1,m)] = \dim[RM(r,m)^{\perp}]$, thus completing the proof of Theorem 9.8.

Appendix. A

Solutions, answers, hints

Chapter 1

1. The principle example of a function we wish to emphasize and illustrate at this point is that of a computation by a computer program.
2. $P(sym) = \left(\frac{\alpha+\beta}{2}\right)^2 \left(3 - 2\left(\frac{\alpha+\beta}{2}\right)\right)$, as shown at the end of Chapter 1.

$$\begin{aligned} P(non-sym) &= P[111 \text{ sent and 2 or 3 errors of type } 1 \to 0 \text{ are made}] \\ &+ P[000 \text{ sent and 2 or 3 errors of type } 0 \to 1 \text{ are made}] \\ &= 0.5\ (\alpha^3 + 3\alpha^2(1-\alpha)) + 0.5(\beta^3 + 3\beta^2(1-\beta)) \end{aligned}$$

   Do some algebra to get $P(sym) - P(non-sym) = \frac{3}{4}(\alpha-\beta)^2(\alpha+\beta-1)$. The sign of this is clearly controlled by the sign of $\alpha+\beta-1$, so we need to know whether $\alpha+\beta > 1$ (an extremely noisy channel!), or, most likely for a realistic channel, $\alpha+\beta < 1$.

   In this latter case $P(sym) - P(non-sym)$ is negative, so the symmetric channel is to be preferred.

3. If one or three errors are made this will cause the received word to have an odd number of 1s, so these errors are detectable. So received words are erroneously accepted if and only if the channel induces two or four errors. The probability of this is

$$\binom{4}{2} p^2(1-p)^2 + p^4 = p^2(6 - 12p + 7p^2).$$

Chapter 2

1. 1011100, 1111111, 0111000.
2. 13 + 49 is encoded as 0001110001101110100100100011100 1001.
   259 ÷ 7 is encoded as 00101010101101100100111010100111000 .

The string can be decoded as

$$\begin{array}{ccccccccc} 10 & - & 8 & , & & 14 \times + , & & 9 \div ? \\ & \uparrow & \uparrow & & \uparrow & * & & * \end{array}$$

The arrows indicate symbols which can be interpreted as 0, −, 'space', respectively, by correcting what is assumed to be a single error in the corresponding 7-bit string. '*' indicates a place where we are reduced to guessing, as 1111111 is a correct encoding of 1111 but this is not one of the 'messages'. Similarly, by assuming one error in the ninth 7-bit string this decodes to + but then the corresponding sum is 14 × + which makes no sense. In the case of the final 1111111 some progress can be made by the exercise of common sense: assuming we are only involved in whole number arithmetic this last symbol could only be 1, 3 or 9. 3 would mean 1111111 contains three errors, but 1 or 9 would both mean four errors – so let's go for 3! Interpreting that doubtful + is more of a problem, and we'd probably be reduced to asking for retransmission.

3. Suppose the transmitted word is $p_1q_2q_3q_4r_5r_6r_7$, and consider in turn the effect on the parity checks of changing p_1, one of the qs, one of the rs. You will see that in all three cases exactly the right parity checks fail in order for the changed bit to be located.

4. 10001110, 11100010, 00110110.

5. 11111111, 11010100, the last word has at least two errors and is not uniquely decodable.

6. With exactly two errors, the errors could be in bits (1,8), (2,5), (3,6) or (4,7).

7. If there are two errors the overall parity check must work, and at least one of the other checks must fail.

   If only A fails the error locations can only be (1,2), (5,8), (3,7), (4,6), and by symmetry there will also be 4 possible pairs if only B or only C fails.

   If A and B but not C fails, the error locations are (5,6), (2,3), (1,7) and (4,8), and similarly when A,C or B,C are the failing pair.

   Finally, if A,B and C fail, the error locations are (1,8), (2,5), (3,6) or (4,7).

   Notice that there are seven possible subsets of {A,B,C} which may fail: {A}, {B}, {C}, {B,C}, {A,C}, {A,B}, {A,B,C}, and we have shown that each is associated with four possibilities for the errors, giving a total of 28 pairs of errors. This provides a check on the reasoning because there are just $\binom{8}{2} = 28$ ways of choosing which two of the eight bits are corrupted.

8. ?0?0111 can only have come from 1000111. xy11001 leads to the parity

checks: (A) $x+1+1$ is even; (B) $x+y+1$ is even; (C) $x+y+1$ is odd, and clearly these are incompatible.

If we assume a single error amongst the recognizable bits, taking the four possibilities for x and y in turn we get:

$x = y = 0$,	B fails ,	implying bit 6 is the error;
$x = y = 1$,	A and B fail ,	implying bit 4 is the error;
$x = 0, y = 1$,	C fails ,	implying bit 7 is the error;
$x = 1, y = 0$,	A and C fail ,	implying bit 3 is the error.

So as in the case of the (8,4) code with two errors, the best we can do is narrow the choice down to four possible words.

9. If the received word is denoted by $p_1q_2q_3q_4r_5r_6r_7$ there are essentially only the following possibilities for where the erasures are:

   p_1 and a q (say p_1 and q_2); p_1 and an r (say p_1 and r_5);
   both qs (say q_2 and q_3); both rs (say r_5 and r_6);
   a q and an r (say q_2 and r_5 or q_2 and r_6).

   For each case the parity checks on A, B and C must be consistent since we know there are no errors. A glance at the diagram for each of the six cases should convince you that in each case the identity of the erasures is uniquely determined.

10. (a) Yes, (b) Yes. Draw the diagrams for erasures in bits 5,6,7 and in bits 4,5,6. The solution in the first case is unique but there are two possibilities in the second.

11. If the (8,4) code is used any word received with three erasures is uniquely recoverable. The basic reason for this is that in those cases in Exercise 10 which led to two possible transmitted codewords, only one of the solutions satisfies the overall parity check.

12. $2^7 = 128$. Each word can be considered to be the result of making seven successive choices: choose first bit, choose second bit, ...; the choices are independent; each one has two outcomes, 0 or 1.

    $2^4 = 16$. In a codeword the last three bits are determined by the first four.

13. The point P could coincide with A or B or be situated on the straight line segment joining them, in which case the correct relationship would be

$$d(A, B) = d(A, P) + d(P, B).$$

14. If the words are of length n, $d(u, w)$ is the sum of the contributions from the n digits. The contribution from the ith digit is 0 if $u_i = w_i$ and 1 if $u_i \neq w_i$. If $u_i = w_i$ then u_i and w_i both agree with v_i or both differ from v_i. So the contribution to $d(u, w)$ is 0 and the contribution to $d(u, v) + d(v, w)$ is 0 or 2, and in both cases the contribution to $d(u, v)$ is $\leq$ that to $d(u, v) + d(v, w)$. On the other hand, if $u_i \neq w_i$ then v_i must

differ from at least one of them, so the contribution to $d(u,v)$ is 1, but that to $d(u,v)+d(v,w)$ is at least one.

So in all cases the ith digit contributes at least as much to $d(u,v)+d(v,w)$ as it does to $d(u,w)$. Hence, by adding all the contributions we get

$$d(u,w) \le d(u,v) + d(v,w).$$

15. q^n – by the same reasoning as in the first part of Exercise 12.
16. (a) at least 4; (b) at least 7.
17. (a) $\alpha = 3,\ \beta = 1$; (b) $\alpha = 4,\ \beta = 2$; (c) $\alpha = 5,\ \beta = 2$
18. The following table gives the distances between all pairs of codewords. From it we see that $d(C) = 2$.

	cbaaa	*bcabc*	*bacbc*	*aabbc*	*acccb*	*cbbab*
cbaaa		4	5	5	5	2
bcabc			2	3	4	5
bacbc				2	4	5
aabbc					4	4
acccb						4

(a) If *cbaaa* is sent and *cbaab* is received (one error), the received word is still closer to *cbaaa* than to any other codeword, so will be correctly decoded.

(b) If *bcabc* is sent and *baabc* received (one error), the received word is at distance one from the codewords *bcabc* and *bacbc* and at a greater distance from all other codewords. Hence, it cannot be decoded.

19. (a) 110110; (b) 101101; (c) not decodable. 110011 has two equally nearest codeword neighbours.
21. Yes. Without lying the parity check bits 5,6 and 7 are irrelevant, so just ask questions 1 to 4 as specified in the lying game.

Chapter 3

1. Let $\boldsymbol{x} = x_1x_2\dots x_7$ and $\boldsymbol{x}' = x'_1x'_2\dots x'_7$ be any two codewords and let $\boldsymbol{y} = y_1y_2\dots y_7 = \boldsymbol{x}+\boldsymbol{x}'$. In order to show that $\boldsymbol{y}$ is a codeword just show that it satisfies all the checks on sets A, B, C.

   For example, for A we have to show that $y_5 = y_1 + y_3 + y_4$ This holds because

$$\begin{aligned} y_5 &= x_5 + x'_5 \\ &= (x_1+x_3+x_4) + (x'_1+x'_3+x'_4) \quad \text{since } \boldsymbol{x} \text{ and } \boldsymbol{x}' \text{ are codewords} \\ &= (x_1+x'_1) + (x_3+x'_3) + (x_4+x'_4) \\ &= y_1 + y_3 + y_4 \end{aligned}$$

2. Let $a = qb + r, a = q'b + r'$ be any two results which satisfy the requirements.

   Then $b(q - q') = r' - r$, so $r' - r$ is a multiple of b, and since $|r' - r|$ is at most $b - 1$ (since $0 \leq r < b, 0 \leq r' < b$), this can only mean $|r' - r| = 0$. Hence $r = r'$ and the equation above shows that $q = q'$ too.

3. (i) and (ii) are trivial.

   For (iii) $$(a \equiv b \bmod m, b \equiv c \bmod m) \Rightarrow (a - b = mk, b - c = ml)$$ $$\Rightarrow \quad a - c = m(k + l) \Rightarrow a \equiv c \bmod m$$

   (iv) Let $a \equiv b \bmod m, c \equiv d \bmod m$. So $a = b + mk, c = d + ml$.

   Adding: $a + c = b + d + m(k + l) \Rightarrow a + c \equiv b + d \bmod m$, and multiplying: $ac = bd + m(kd + mkl + bl) \Rightarrow ac = bd \bmod m$.

4. Put c, d equal to a, b respectively in (iv) to get $a^2 \equiv b^2 \bmod m$. Using this with $a \equiv b \bmod m$ and applying (iv) again, get $a^3 \equiv b^3 \bmod m \ldots$ and so on.

   Working mod 41, $2^{20} = (2^5)^4 \equiv (-9)^4 = 81^2 \equiv (-1)^2 = 1$.
   That is, $41 | 2^{20} - 1$.

   We need 2^{50} and 41^{65} worked out mod 7:
   $2^{50} = (2^3)^{16} 2^2 \equiv 1^{16} 2^2 = 4$, and
   $41^{65} \equiv (-1)^{65} = -1 \equiv 6$.

   Working mod 4,

$$\begin{aligned}
\Sigma_{i=1}^{100} i^5 &= (1^5 + 2^5 + 3^5 + 4^5) + (5^5 + 6^5 + 7^5 + 8^5) + \ldots + 100^5 \\
&\equiv (1^5 + 2^5 + 3^5 + 4^5) + (1^5 + 2^5 + 3^5 + 4^5) + \ldots + 4^5 \\
&= 25 \times (1^5 + 2^5 + 3^5 + 4^5) \\
&\equiv 25 \times (1^5 + 2^5 + (-1)^5 + 0) \\
&\equiv 25 \times (1 + 0 - 1 + 0) = 0.
\end{aligned}$$

5. (a) a counter-example is $a = 1, b = 5, m = 4, c = 2$.

   (b) a counter-example is $a = 2, b = 4, m = 4$.

6. $n \equiv 0, 1, 2, 3$ or 4 mod 5. The corresponding values of n^4 mod 5 are $0, 1, 16, 81, 256 \equiv 0, 1, 1, 1, 1$ respectively.

7.
$$\begin{aligned}
n &\equiv 0,\ 1,\ 2,\ 3,\ 4 \text{ or } 5 && \text{mod 6, so} \\
n + 1 &\equiv 1,\ 2,\ 3,\ 4,\ 5, \text{ or } 0 && \text{respectively and} \\
2n + 1 &\equiv 1,\ 3,\ 5,\ 1,\ 3 \text{ or } 5 && \text{respectively,}
\end{aligned}$$

   and it is now easy to check that $n(n + 1)(2n + 1) \equiv 0$ mod 6 in all cases.

8.
$$\begin{aligned}
&\text{(a)} \quad (a|b, c|d) &&\Rightarrow \quad (b = ak, d = cl) &&\Rightarrow \quad bd = ac.kl \\
&\text{(b)} \quad (a|b, b|a) &&\Rightarrow \quad (b = ak, a = bl) &&\Rightarrow \quad ab = ab.nl \\
& &&\Rightarrow \quad ab = 0 \text{ or } kl = 1 &&
\end{aligned}$$

   Now $ab = 0 \Rightarrow a = 0$ or $b = 0 \Rightarrow a, b$ are *both* zero, otherwise $a|b$ or $b|a$ would be false.

   Hence $a = b = 0$ or $kl = 1$.

In the former case, obviously $a = b$.

In the latter $k = l = \pm 1$, so $a = b$ or $a = -b$.

$$\text{(c)}\ (a|b, b \neq 0) \Rightarrow (b = ka, b \neq 0) \Rightarrow (|b| = |k||a|, b \neq 0)$$
$$\Rightarrow (|b| = |k||a|, k \neq 0) \Rightarrow (|a| = \tfrac{|b|}{|k|} \leq |b|)$$
$$\text{(d)}\ (a|b, a|c) \Rightarrow (b = ka, c = ln) \Rightarrow bx + cy = a(kx + ly)$$

9. $a \equiv 0, 1$ or $2 \bmod 3$. $a \equiv 0 \Rightarrow 3|a, a \equiv 1 \Rightarrow 3|a + 2, a \equiv 2 \Rightarrow 3|a + 4$.
10. $a^2 - 1 = (a-1)(a+1)$. $2 \not| a$ so $a-1, a+1$ are consecutive even integers. This implies both are divisible by 2 and one them is divisible by 4.

    Hence $8|(a-1)(a+1)$ (1)

    $3 \not| a$, and since one of three consective integers $a-1, a, a+1$ must be divisible by 3 we have $3|(a-1)(a+1)$ (2)

    From (1) and (2) it follows that $24|a^2 - 1$.
11. $\gcd(a, b) = ax + by, a = kc, b = lc$, so $\gcd(a, b) = c(kx + ly)$.
12. Let d be any common divisor of $2n + 1$ and $n^2 + 3n + 1$.

    Then $d| - n(2n + 1) + 2(n^2 + 3n + 1)$ by Exercise 8 (iv).

    That is $d|5n + 2$.

    So $d|2n + 1$ and $d|5n + 2$, so by applying 8 (iv) again we get

$$d|5(2n + 1) - 2(5n + 2).$$

    That is, $d|1$.
13. Suppose 3 consecutive steps in Euclid's algorithm are:

    (a) $r_i = qr_{i+1} + r_{i+2}$
    (b) $r_{i+1} = q'r_{i+2} + r_{i+3}$
    (c) $r_{i+2} = q''r_{i+3} + r_{i+4}$

    If the difference between r_{i+2} and r_{i+3} is small (only a fraction of r_{i+2}) then step 3 will have $q'' = 1$, so $r_{i+2} = r_{i+3} + r_{i+4}$, so r_{i+4} will be small, and the jump from r_{i+3} to r_{i+4} is big.
14. Hint: show that if $T_{24,42}$ contains the number x it must also contain every multiple of x. If you can't get any further be patient until theorem 3.4.
15. By applying Euclid and reversing the steps, you should obtain

$$\gcd(1729, 703) = 19 = 32 \times 703 \quad - \quad 13 \times 1729,$$

    so a solution of $1729x + 703y = 19$ is $x = 32, y = -13$. Further solutions can be obtained by adding any multiple of 703 to x and subtracting the same multiple of 1729 from y.

    Using the same method for $25x + 35y = 15$ we obtain

$$\gcd(25, 35) = 5 = 3 \times 25 \quad - \quad 2 \times 35,$$

and to get 15 on the right hand side we need to multiply by 3. So $x = 9, y = -6$ is a solution and further solutions are $x = 9 + 35k, y = -6 - 25k$. But this time we don't get *all* solutions in this way. Why?

16. $$\begin{aligned} \gcd(a,b) = d \Rightarrow \quad & d = ax + by \Rightarrow 1 = (\tfrac{a}{d})x + (\tfrac{b}{d})y \\ \Rightarrow \quad & \gcd(\tfrac{a}{d}, \tfrac{b}{d})|1 \text{ by Theorem 3.4} \\ \Rightarrow \quad & \gcd(\tfrac{a}{d}, \tfrac{b}{d}) = 1 \end{aligned}$$

17. If p is prime and $p|bc$ and $p \not| b$, then $p|c$.

18. $$\begin{aligned} a(x+lm) - c &= (ax - c) + alm \\ &= km + alm, (\text{because } ax \equiv c \bmod m) \\ &\equiv 0 \bmod m. \end{aligned}$$

so $a(x + lm) \equiv c \bmod m$.

19.(a) $(a \equiv b \bmod n, m|n) \Rightarrow (a - b = kn, n = lm) \Rightarrow a - b = klm.$

(b) $a \equiv b \bmod m \Rightarrow a - b = km \Rightarrow ca - cb = kcm.$

(c) $a \equiv b \bmod m$ means $a = mq + b$ so the required result is just Theorem 3.2.

20. $0, 1, 2, 2^2, 2^3, \ldots 2^9 \equiv 0, 1, 2, 4, 8, 5, 10, 9, 7, 3, 6$ respectively.

$$1^2 \equiv 10^2$$

The list $ka_1, ka_2, \ldots, ka_m$ has m members so we only need to show that they are all different mod m. Suppose $ka_i \equiv ka_j \bmod m$ with $a_i \neq a_j$. Then $a_i \equiv a_j \bmod m$ by the gcd condition, which implies $a_i = a_j$ by the given property of the as.

21.(a) $$\begin{aligned} & 4x \equiv 5 \bmod 7 \\ \Leftrightarrow \quad & 4x \equiv 12 \bmod 7 \\ \Leftrightarrow \quad & x \equiv 3 \bmod 7 \end{aligned}$$

(b) $$\begin{aligned} & 8x \equiv 12 \bmod 19 \\ \Leftrightarrow \quad & 2x \equiv 3 \bmod 19 \\ \Leftrightarrow \quad & 2x \equiv 22 \bmod 19 \\ \Leftrightarrow \quad & x \equiv 11 \bmod 19 \end{aligned}$$

(c) $12x \equiv 3 \bmod 4$ has no solutions because $\gcd(12, 4) = 4$ and $4 \not| 3$.

(d) $$\begin{aligned} & 45x \equiv 75 \bmod 100 \\ \Leftrightarrow \quad & 3x \equiv 5 \bmod 20 \\ \Leftrightarrow \quad & 3x \equiv 45 \bmod 20 \\ \Leftrightarrow \quad & x \equiv 15 \bmod 20 \end{aligned}$$

(e) $$\begin{aligned} & 111x \equiv 112 \bmod 113 \\ \Leftrightarrow \quad & -2x \equiv -1 \bmod 113 \\ \Leftrightarrow \quad & -2x \equiv 112 \bmod 113 \\ \Leftrightarrow \quad & x \equiv -56 \\ & \equiv 57 \bmod 113 \end{aligned}$$

(f)
$$\begin{aligned} & 140x & \equiv & \; 133 \bmod 301 \\ \Leftrightarrow & \; 20x & \equiv & \; 19 \bmod 43 \\ \Leftrightarrow & \; 20x & \equiv & \; 105 \bmod 43 \\ \Leftrightarrow & \; 4x & \equiv & \; 21 \bmod 43 \\ \Leftrightarrow & \; 4x & \equiv & \; 64 \bmod 43 \\ \Leftrightarrow & \; x & \equiv & \; 16 \bmod 43 \end{aligned}$$

22. Properties (iii) and (iv) of theorem 3.1 and Exercise 4.

23. (a)
$$\begin{aligned} 478034 &= 4 + 3.10 + 0.10^2 + 8.10^3 + 7.10^4 + 4.10^5 \\ &\equiv 4 + 3(-1) + 0(-1)^2 + 8(-1)^3 + 7(-1)^4 + 4(-1)^5 \\ &= 4 - 3 + 0 - 8 + 7 - 4 \end{aligned}$$

(b) $10 \equiv -3 \bmod 13$ so $10^3 \equiv -27 \equiv -1 \bmod 13$.

Hence
$$\begin{aligned} 2398047812 &= 812 + 47.10^3 + 398.(10^3)^2 + 2.(10^3)^3 \\ &= 812 - 47 + 398 - 2 \bmod 13 \end{aligned}$$

24. Suppose the individual digits of n are $a_{2k-1}a_{2k-2}\ldots a_2a_1a_0$, so that $n = a_0 + 10a_1 + 10^2a_2 + \ldots + 10^{2k-1}a_{2k-1}$.

Then $m = a_1 + 10a_2 + 10^2a_3 + \ldots + 10^{2k-1}a_0$. so

$$\begin{aligned} n + m &= a_0(1 + 10^{2k-1}) + a_1(1 + 10) + a_2(10 + 10^2) \\ &\quad + \ldots + a_{2k-1}(10^{2k-2} + 10^{2k-1}). \end{aligned}$$

Now $10 \equiv -1 \bmod 11$ so each bracket in the above expression is $\equiv 0 \bmod 11$. Hence $n + m \equiv 0 \bmod 11$. $n^2 - m^2 = (n - m)(n + m)$ so it remains to prove $a|n - m$.

$$\begin{aligned} n - m &= a_0(1 - 10^{2k-1}) + a_1(10 - 1) + a_2(10^2 - 10) \\ &\quad + \ldots + a_{2k-1}(10^{2k-1} - 10^{2k-2}), \end{aligned}$$

and because $10 \equiv 1 \bmod 9$, each of these brackets is $\equiv 0 \bmod 9$.

25. If we write down the Fibonacci sequence replacing each term a_i by its least residue b_i mod 21, then the sequence $\langle b_i \rangle$ will satisfy $b_1 = b_2 = 1, b_{i+2} \equiv b_{i+1} + b_i \bmod 21$ for all $i \geq 1$.

Hence $\langle b_i \rangle$ is 1, 1, 2, 3, 5, 8, 13, 0, 13, 13, 5, 18, 2, 20, 1, 0, 1, 1, ..., so the complete sequence is just an endless repetition of the first 16 terms. In this block b_8 and b_{16} are zero and there are no terms equal to 7 or 14, Hence, in $\langle a_i \rangle$ every 8th term is divisible by 21 and there are no other terms divisible by 7.

26. The trick is to work modulo 4. $999 \equiv 3 \bmod 4$, but x^2 and y^2 can only be 0 or 1 mod 4, so $x^2 + y^2$ can only be congruent to 0, 1 or 2.

27. a composite $\Rightarrow a = xy$ with $x > 1, y > 1, x < a, y < a$. If x and y are different, then clearly both occur in the list $1, 2, 3, \ldots, a-1$, so $xy|(a-1)!$

If $a = x^2$ and $a \geq 6$, then $x > 2$, so $2x < x.x = a$. Therefore x and $2x$ both occur in the list $1, 2, 3, \ldots, a - 1$. So $x.2x|(a - 1)!$, so $x^2|(a - 1)!$.

28.(a)
$$\begin{aligned} \gcd(a,b) = p &\Rightarrow a = pk, b = pl, \gcd(k,l) = 1 \\ &\Rightarrow \gcd(a^2, bp) = \gcd(p^2k^2, p^2l) = p^2\gcd(k^2, l) \qquad (1) \\ \gcd(k,l) = 1 &\Rightarrow kx + ly = 1 \Rightarrow (kx+ly)^2 = 1 \\ &\Rightarrow k^2(x^2) + l(ly^2 + 2kxy) = 1 \Rightarrow \gcd(k^2, l) = 1. \end{aligned}$$

so from (1), $\gcd(a^2, bp) = p^2$.

(b) This is false:
$$\begin{aligned} & (\gcd(a,p^2) = p, \gcd(b,p^2) = p^2) \\ \Rightarrow\ & (a = pk, p \nmid k, b = lp^2) \\ \Rightarrow\ & \gcd(ab, p^4) = \gcd(p^3lk, p^4), \end{aligned}$$

which could be p^4 as there is no reason why l shouldn't be a multiple of p. A simple counter-example is $a = 2, p = 2, b = 4$.

(c) Also false by similar reasoning.

(d) Let $d = \gcd(a,b)$, so $a = kd, b = ld$ for some $k, l \geq 1$. Then

$$a^2 + b^2 = k^2d^2 + l^2d^2 = d^2(k^2 + l^2) = p^2.$$

Since p is prime the Fundamental Theorem of Arithmetic implies that $d^2 = 1, k^2 + l^2 = p$ or $d^2 = p, k^2 + l^2 = p$, or $d^2 = p^2, k^2 + l^2 = 1$. The second of these is impossible since a prime cannot be a square, and so is the third since $k^2, l^2 \geq 1$, so $k^2 + l^2$ can't be 1.

29. The method clearly yields *a* common divisor, but if the two numbers had prime factorizations different from those given there would be no guarantee that the same method applied to the new factorizations would yield the same common divisor.

30. If i^{th} member of $C_1 = j^{th}$ member of C_2, then (if $i \geq j$), the first member of C_2 would be the same as the $i - (j-1)^{th}$ member of C_1, which contradicts the rule by which C_2 was constructed. A similar contradiction is obtained if $i < j$ because then the first member of C_2 would be the $(p - j + i + 1)^{th}$ member of C_1.

31. $195 = 5 \times 13 \times 3$. Use the corollary to Fermat's theorem to establish the congruences $a^{195} \equiv a$ to each of the moduli 5, 13, 3. You will then have shown that $a^{195} - a$ is a multiple of 5, 13 and 3, so it is a multiple of $5 \times 13 \times 3$.

32. Modulo 31:
$$\begin{aligned} 99^{101} &\equiv 6^{101} = (6^{30})^3, 6^{11} &&\equiv 6^{11} = 36^5.6 \\ &\equiv 5^5.6 = 30.5^4 &&\equiv (-1).25^2 \\ &\equiv (-1)(-6)^2 \equiv -36 &&\equiv -5 \equiv 26. \end{aligned}$$

33. $341 = 11 \times 31$, so is not prime. But $2^{340} = (2^5)^{68} \equiv (-1)^{68} \bmod 11 = 1$ and $2^{340} = (2^5)^{68} \equiv 1^{68} \bmod 31 = 1$. Hence $2^{340} \equiv 1 \bmod 341$.

34. Clearly $10^1 \equiv 4 \bmod 6$, and *if* $10^n \equiv 4 \bmod 6$, then $10^{n+1} = 10(10^n) \equiv 40 \equiv 4 \bmod 6$, so by induction the required result holds for all $n \geq 1$.

Let m be the larger of m and n. Then $m = 6k + n$ for some $k \geq 0$, and working modulo 7 we have $10^m = 10^n(10^6)^k \equiv 10^n$ by Fermat's theorem.

By combining these results we have $10, 10^2, 10^3, \ldots, 10^{10}$ all $\equiv 4 \bmod 6$, so $10^{10}, 10^{(10^2)}, 10^{(10^3)}, \ldots, 10^{(10^{10})}$ are all $\equiv 10^{10} \bmod 7$.

Now $10^{10} \equiv 3^{10} = (3^2)^5 \equiv 2^5 = 32 \equiv 4 \bmod 7$.

So the given expression $\equiv 10 \times 4 \equiv 5 \bmod 7$.

35. Applying the same method as for the normal pack you should find that n shuffles suffice where n has to be a positive solution of $2^n \equiv 1 \bmod 55$. The difference is that 55 is not prime so regard this congruence as equivalent to the pair, $2^n \equiv 1 \bmod 5$ and $2^n \equiv 1 \bmod 11$. By Fermat's theorem $n = 4$ satisfies the first, and $n = 10$ satisfies the second. So their lcm, 20, will satisfy both, and as in the analysis for the normal pack, any smaller n which satisfies both will be a factor of 20. It is easy to check that 1, 2, 4, 5, 10 don't work, so 20 shuffles are necessary.

36. Using the hint, number the positions in the pack from 0 to 51 rather than 1 to 52. Then you should find that the result of a shuffle is to send the card originally in position x to position $2x$ *mod 51*. So the equation to be solved this time is $2^n \equiv 1 \bmod 51$, and $n = 8$ is the smallest positive solution.

37.
$$a' + b' = (a + km) + (b + lm) = a + b + (k + l)m \equiv a + b \equiv c \equiv c' \bmod m$$

and

$$a'b' = (a + km)(b + lm) = ab + m(kb + la + mkl) \equiv ab \equiv c \equiv c' \bmod m$$

38.

x :	1	2	3	4	5	6
x^{-1}:	1	4	5	2	3	6

$32x \equiv 40 \bmod 7$ has only one solution mod 7 since $\gcd(32, 7) = 1$.

$$32 \equiv 4 \bmod 7 \text{ and } 4^{-1} = 2.$$

So multiplying by 2 we get

$$64x \equiv 80 \bmod 7,$$

that is $x \equiv 80 \equiv 3 \bmod 7$.

Chapter 4

1. For the 7-bit code $M = 16, q = 2, n = 7, d = 3$, so $q^{n-d+1} = 2^5 \neq 16$ so this code is not MDS. For the 8-bit code $M = 16, q = 2, n = 8$, so to be MDS d would have to satisfy $16 = 2^{9-d}$. That is, $d = 5$. But this is impossible because the 7-bit code contains words $\mathbf{c}, \mathbf{c}'$ for which $d(\mathbf{c}, \mathbf{c}') = 3$, so the corresponding pairs of words in the 8-bit code differ in at most 4 places.

2. Consider a channel which induces at most t errors. If the code satisfies the geometric condition and the transmitted codeword $\boldsymbol{u}$ is received as $\boldsymbol{v}$, then $\boldsymbol{v} \in S(\boldsymbol{u}, t)$. But $\boldsymbol{v}$ is not in any of the other codeword-

centred spheres, so will be correctly decoded as $\boldsymbol{u}$. If the condition is not satisfied, let $\boldsymbol{v} \in S(\boldsymbol{u},t) \cap S(\boldsymbol{u}',t)$ with $d(\boldsymbol{u}',\boldsymbol{v}) \leq d(\boldsymbol{u},\boldsymbol{v}) \leq t$. Suppose $\boldsymbol{u}$ is sent and $\boldsymbol{v}$ received. If $d(\boldsymbol{u}',\boldsymbol{v}) = d(\boldsymbol{u},\boldsymbol{v})$ then the decoder cannot unequivocally decode $\boldsymbol{v}$ as $\boldsymbol{u}$, and if $d(\boldsymbol{u}',\boldsymbol{v}) < d(\boldsymbol{u},\boldsymbol{v})$ then it will definitely *not* decode $\boldsymbol{v}$ as $\boldsymbol{u}$.

3. With $q = 3, d \geq 5, n = 10$ the Hamming bound gives $M \leq 294$, so the answer is no.

5. For these codes $t = 1, d = 3$, so the result (4.7) is $q^2 > 1 + n(q-1)$ which implies $n < q + 1$.

6. If not, the process would not be complete because one of the uncovered words, which would have distance $\geq d$ from all the sphere centres, could be added to C.

7.(a) $6808 \leq M \leq 10^6$ with Singleton being the better of the upper bounds.

(b) $5 \leq M \leq 22$ with Hamming being the better upper bound.

(c) $2 \leq M \leq 9$ with Singleton being the better upper bound.

The Hamming bound depends on q, n and the error correcting capability. (b) and (c) have the same q and n, and the error correcting capability is 1 for both $d = 3$ and $d = 4$.

To improve the lower bound of 5 in (b) we need only construct a (5,6,3) code over Z_3. A bit of intelligent trial and error should suffice. One example is $\{00000, 11100, 22200,$
$01210, 00121, 22111\}$.

8.

$$C = \begin{array}{c} aadcca \\ adcacd \\ cdabaa \\ dcbdbc \end{array} \quad \underset{\text{positions } 1,3}{\overset{\text{switch}}{\longrightarrow}} \quad \begin{array}{c} daacca \\ cdaacd \\ adcbaa \\ bcddbc \end{array} \qquad \text{(this achieves (i))}$$

$$\underset{\text{positions } 4,6}{\overset{\text{switch}}{\longrightarrow}} \quad \begin{array}{c} daaacc \\ cdadca \\ adcaab \\ bcdcbd \end{array} \quad \underset{\substack{b \text{ by } a, c \text{ by } d, \\ d \text{ by } b \text{ in last} \\ \text{place}}}{\overset{\text{replace } a \text{ by } c}{\longrightarrow}} \quad \begin{array}{c} daaacd \\ cdadcc \\ adcaaa \\ bcdcbb \end{array} \qquad \text{(this achieves (ii))}$$

$$\underset{\text{in positon } 3}{\overset{\text{switch } a \text{ and } b}{\longrightarrow}} \quad \begin{array}{c} dabacd \\ cdbdcc \\ adcaaa \\ bcdcbb \end{array} \quad = C'.$$

9. $C_1 = \{ddbba, bcaad, bbabb, dccbd, adcdd, ccbac\}$

$C_2 = \{ddbca, bcaac, bbacd, dcccc, adcbc, ccbab\}$

10. $A_2(5,3) = 4, \qquad A_2(9,5) = 6.$

11. Let C be a binary (n, M, d) code. Following the hint, let S be a set of M' words of C, beginning with the same symbol, with $M' \geq \frac{M}{2}$.

    Let S' be the set of words of length $n - 1$ formed by deleting this first symbol from the words of S.

    Regarding S as a subcode of C, it is clear that $d(S) = d' \geq d$. Hence $d(S') = d' \geq d$, and S' is the required code.

12. The codes $C = \{000, 111, 010\}$ and $C' = \{011, 010, 001\}$ both have minimum distance 1. C' has one position in which all the symbols are the same, and all the equivalence operations preserve this property. C does not have the property, so no sequence of equivalence operations applied to C' can produce C. Hence C and C' are not equivalent.

13. C has the same symbol in position 8 of all words. C' has no such position. See solution of Exercise 12.

15. Suppose C is perfect with $d(C) = 2x$ for some positive integer x. Let $\mathbf{c}_1, \mathbf{c}_2$ be codewords with $d(\mathbf{c}_1, \mathbf{c}_2) = 2x$, and let T be the set of positions at which $\mathbf{c}_1$ and $\mathbf{c}_2$ differ, so $|T| = 2x$. Take $\mathbf{c}_1$ and x of these positions. Change the symbols in these positions to the corresponding symbol of $\mathbf{c}_2$, and let $\boldsymbol{r}$ be the resulting word.

    Then $d(\mathbf{c}_1, \boldsymbol{r}) = d(\mathbf{c}_2, \boldsymbol{r}) = x$ and since C is a perfect $(x - 1)$-error correcting code $\boldsymbol{r}$ must be in some sphere $S(\mathbf{c}, x - 1)$ with $\mathbf{c} \neq \mathbf{c}_1, \mathbf{c} \neq \mathbf{c}_2, \mathbf{c} \in C$. Then using the triangle inequality we have

    $$d(\mathbf{c}_1, \mathbf{c}) \leq d(\mathbf{c}_1, \boldsymbol{r}) + d(\boldsymbol{r}, \mathbf{c}) \leq x + (x - 1) = 2x - 1,$$

    which is impossible since $d(C) = 2x$.

16. Let $\mathbf{c}$ be sent and $\boldsymbol{r}$ received. $\boldsymbol{r} \in S(\mathbf{c}', t)$ for some unique $\mathbf{c}' \in C$ by Theorem 4.7, and $d(\mathbf{c}, \boldsymbol{r}) > t$. Hence nearest neighbour decoding must decode $\boldsymbol{r}$ to $\mathbf{c}'$, not $\mathbf{c}$.

17. Let x_i, y_i be the ith place symbol in $\boldsymbol{x}$ and $\boldsymbol{y}$ respectively, and let $w_i(\boldsymbol{x})$ denote the contribution (0 or 1) to $w(\boldsymbol{x})$ from this place. The table below lists the relevant quantities for each of the x_i, y_i pairs.

x_i	y_i	$w_i(\boldsymbol{x}+\boldsymbol{y})$	$w_i(\boldsymbol{x}) + w_i(\boldsymbol{y})$	$(\boldsymbol{x} \odot \boldsymbol{y})_i$
0	0	0	0	0
0	1	1	1	0
0	2	1	1	0
1	1	1	2	1
1	2	0	2	2
2	2	1	2	1

The entries in the final column are precisely the values of $w_i(\boldsymbol{x})+w_i(\boldsymbol{y})-$

$w_i(\boldsymbol{x}+\boldsymbol{y})$, so $f(\boldsymbol{x}\odot\boldsymbol{y})$ must contribute 2 for every place in which $\boldsymbol{x}\odot\boldsymbol{y}$ has 2, 1 for each place in which it has 1, and 0 for the rest.

$f(\boldsymbol{a})$ = number of 1s + 2 × number of 2s in $\boldsymbol{a}$ will achieve this.

18. If each word $\boldsymbol{x}$ in Z_2^8 is interpreted as a 'received word', then the decoding algorithm of Chapter 2 shows how to find a codeword $\mathbf{c}$ (not necessarily unique) such that $d(\boldsymbol{x},\mathbf{c})\leq 2$.

19. Using the first equality of Theorem 4.7,

$$d(\boldsymbol{a}+\boldsymbol{b},\boldsymbol{x}+\boldsymbol{y})=w((\boldsymbol{a}+\boldsymbol{b})+(\boldsymbol{x}+\boldsymbol{y}))=w((\boldsymbol{a}+\boldsymbol{x})+(\boldsymbol{b}+\boldsymbol{y}))=d(\boldsymbol{a}+\boldsymbol{x},\boldsymbol{b}+\boldsymbol{y})$$

20. To each ordered pair of codewords $(\boldsymbol{u},\boldsymbol{v})$ of $Z_2^n\times C$ there corresponds a unique codeword $\boldsymbol{u}|\boldsymbol{u}+\boldsymbol{v}|f(\boldsymbol{u})$ of D, and this correspondence is one to one since if $\boldsymbol{u}\neq\boldsymbol{u}'$ then $\boldsymbol{u}|\boldsymbol{u}+\boldsymbol{v}|f(\boldsymbol{u})\neq\boldsymbol{u}'|\boldsymbol{u}'+\boldsymbol{v}'|f(\boldsymbol{u}')$, and if $\boldsymbol{u}=\boldsymbol{u}'$ and $\boldsymbol{v}\neq\boldsymbol{v}'$, then $\boldsymbol{u}+\boldsymbol{v}\neq\boldsymbol{u}'+\boldsymbol{v}'$, so again $\boldsymbol{u}|\boldsymbol{u}+\boldsymbol{v}|f(\boldsymbol{u})\neq\boldsymbol{u}'|\boldsymbol{u}'+\boldsymbol{v}'|f(\boldsymbol{u}')$.

    Hence $|D|=|Z_2^n\times C|=|Z_2^n|\times|C|=2^n.2^{n-m}=2^{2n-m}$.

21. Let $\mathbf{c},\mathbf{c}'$ be distinct codewords of D. In order to show that $d(\mathbf{c},\mathbf{c}')\geq 3$ we split into 3 cases.

    Case 1: $\mathbf{c}=\boldsymbol{u}|\boldsymbol{u}+\boldsymbol{v}|f(\boldsymbol{u}),\mathbf{c}'=\boldsymbol{u}|\boldsymbol{u}+\boldsymbol{v}'|f(\boldsymbol{u}),\boldsymbol{v}\neq\boldsymbol{v}'$.
    In this case $d(\mathbf{c},\mathbf{c}')=d(\boldsymbol{u}+\boldsymbol{v},\boldsymbol{u}+\boldsymbol{v}')=d(\boldsymbol{u}+\boldsymbol{u},\boldsymbol{v}+\boldsymbol{v}')$
    $=d(\boldsymbol{o},\boldsymbol{v}+\boldsymbol{v}')=w(\boldsymbol{v}+\boldsymbol{v}')=d(\boldsymbol{v},\boldsymbol{v}')\geq 3$, since $d(C)=3$.

    Case 2: $\mathbf{c}=\boldsymbol{u}|\boldsymbol{u}+\boldsymbol{v}|f(\boldsymbol{u}),\mathbf{c}'=\boldsymbol{u}'|\boldsymbol{u}'+\boldsymbol{v}|f(\boldsymbol{u}'),\boldsymbol{u}\neq\boldsymbol{u}'$.
    Using the same trick as for case 1, $d(\boldsymbol{u}+\boldsymbol{v},\boldsymbol{u}'+\boldsymbol{v})=d(\boldsymbol{u},\boldsymbol{u}')$.
    So if $d(\boldsymbol{u},\boldsymbol{u}')\geq 2$, then $d(\mathbf{c},\mathbf{c}')\geq 2+2+d(f(\boldsymbol{u}),f(\boldsymbol{u}'))\geq 4$.
    If $d(\boldsymbol{u},\boldsymbol{u}')$ is only 1, then $d(\boldsymbol{u}+\boldsymbol{v},\boldsymbol{u}'+\boldsymbol{v})$ is also 1.
    and $d(\mathbf{c},\mathbf{c}')=1+1+d(f(\boldsymbol{u}),f(\boldsymbol{u}'))$.
    But by using the second part of Theorem 4.7,
    $1=d(\boldsymbol{u},\boldsymbol{u}')=w(\boldsymbol{u})+w(\boldsymbol{u}')-2w(\boldsymbol{u}\odot\boldsymbol{u}')$,
    so $\boldsymbol{u}$ and $\boldsymbol{u}'$ must have opposite parity, and $d(f(\boldsymbol{u}),f(\boldsymbol{u}'))=1$.
    Hence $d(\mathbf{c},\mathbf{c}')=1+1+1=3$.

    Case 3: $\mathbf{c}=\boldsymbol{u}|\boldsymbol{u}+\boldsymbol{v}|f(\boldsymbol{u}),\mathbf{c}'=\boldsymbol{u}'|\boldsymbol{u}'+\boldsymbol{v}'|f(\boldsymbol{u}'),\boldsymbol{u}\neq\boldsymbol{u}',\boldsymbol{v}\neq\boldsymbol{v}'$.
    The only potential problem values here are $d(\boldsymbol{u},\boldsymbol{u}')=2$,
    $d(\boldsymbol{u}+\boldsymbol{v},\boldsymbol{u}'+\boldsymbol{v}')=0$
    and $d(\boldsymbol{u},\boldsymbol{u}')=1,d(\boldsymbol{u}+\boldsymbol{v},\boldsymbol{u}'+\boldsymbol{v}')\leq 1$.
    In the first of these $\boldsymbol{u}+\boldsymbol{v}=\boldsymbol{u}'+\boldsymbol{v}'$, so $\boldsymbol{u}+\boldsymbol{u}'=\boldsymbol{v}+\boldsymbol{v}'$
    (remember $-\boldsymbol{u}=\boldsymbol{u}$ in binary!)
    But since $d(\boldsymbol{u},\boldsymbol{u}')=2,w(\boldsymbol{u}+\boldsymbol{u}')=2$, but $w(\boldsymbol{v}+\boldsymbol{v}')=2$,
    ie $d(\boldsymbol{v},\boldsymbol{v}')=2$, which contradicts $d(C)=3$.

    The second also cannot occur because $d(\boldsymbol{u}+\boldsymbol{v},\boldsymbol{u}'+\boldsymbol{v}')=d(\boldsymbol{u}+\boldsymbol{u}',\boldsymbol{v}+\boldsymbol{v}')$ and $\boldsymbol{u}+\boldsymbol{u}'$ has weight 1, whereas $w(\boldsymbol{v}+\boldsymbol{v}')=d(\boldsymbol{v},\boldsymbol{v}')\geq 3$, so $\boldsymbol{u}+\boldsymbol{u}'$ and $\boldsymbol{v}+\boldsymbol{v}'$ must differ in at least two places.

22. Proving perfection is now easy!

D is a $(2n+1, 2^{2n-m}, 3)$ code, so checking for equality in the sphere packing bound,

$$2^{2n+1}\left[\sum_{i=0}^{1}\binom{2n+1}{i}\right]^{-1} = 2^{2n+1}[1+2n+1]^{-1}$$

$$= 2^{2n+1}[2(2^m-1)+2]^{-1} = 2^{2n+1}2^{-m-1} = 2^{2n-m} = |D|.$$

23. x can be written as $a+b$ with $a \in Z$ and $0 \leq b < 1$.

If $0 \leq b < 0.5$	$2[x] = [2x] < 2[x]+1.$
If $0.5 \leq b < 1$	$2[x] < [2x] = 2[x]+1.$
so for all x,	$2[x] \leq [2x] \leq 2[x]+1$.

24. The argument still works up to the point (case 1), $(M^2-M)d \leq \frac{nM^2}{2}$. This implies $2(M-1)d \leq nM$, so $M(2d-n) \leq 2d$.

But $2d-n$ is now negative, so from the last inequality we get $M \geq \frac{2d}{2d-n}$.

So M is bigger than something negative! – another of those results in the 'true – but not a great deal of use' category. Case 2 is similarly uninformative.

25.

n	Singleton estimate	Hamming estimate	Plotkin estimate	Exact value
13	128	21	14	8
12	64	13	7	4
11	32	8	4	4
10	16	5	3	2

Chapter 5

[If you know more linear algebra than what is in section 5.3 you may well find slicker methods to solve some of these exercises.]

1. $\boldsymbol{u}+\boldsymbol{v} = 0110011 = \boldsymbol{u}-\boldsymbol{v}, \quad -\boldsymbol{u} = \boldsymbol{u}, \ \alpha\boldsymbol{v} = \boldsymbol{0}, \ \beta\boldsymbol{v} = \boldsymbol{v}.$
   $\boldsymbol{u}-\boldsymbol{v} = 1131013, \alpha\boldsymbol{u}+\beta\boldsymbol{v} = 0142011.$

2. (i) $\Rightarrow \alpha\mathbf{c},\ \beta\mathbf{c} \in C$, then (ii) $\Rightarrow \alpha\mathbf{c}+\beta\mathbf{c}' \in C$. Conversely, $\alpha\mathbf{c}+\beta\mathbf{c}' \in C$ for all α and $\beta \Rightarrow \alpha\mathbf{c} \in C$ for all α (put $\beta = 0$) and $\mathbf{c}+\mathbf{c}' \in C$ (put $\alpha = \beta = 1$).

3. Let $\mathbf{c} \in C$. Then if $\alpha = 1$ clearly $\alpha\mathbf{c} \in C$, and if $\alpha = 0, \alpha\mathbf{c} = \mathbf{0}$ and $\mathbf{0} \in C$ because $\mathbf{c}+\mathbf{c} \in C$ by (i) and $\mathbf{c}+\mathbf{c} = \mathbf{0}$.

4. $w(\mathbf{c}_1-\mathbf{c}_2)$ = number of places in which $\mathbf{c}_1-\mathbf{c}_2$ has a non-zero symbol.
   = number of places in which $\mathbf{c}_1, \ \mathbf{c}_2$ differ
   $= d(\mathbf{c}_1, \mathbf{c}_2)$.

5. No, 1; Yes, 1; No, 1; Yes, 3; Yes, 2; No, 2.

6. Let C be a linear binary code in which not every codeword begins with 0. Let C_0 be the set of codewords which do start with 0 (note $\mathbf{0} \in C$ so $C_0 \neq \phi$), and C_1 the set of codewords which start with 1, so

$$C_0 \cap C_1 = \phi.$$

Let $y_1 \in C_1$ be chosen.

Show that $\boldsymbol{y}_1 + \boldsymbol{y}(\boldsymbol{y} \in C_1)$ are distinct words of C_0 so $|C_1| \leq |C_0|$, and $\boldsymbol{x} + \boldsymbol{y}_1(\boldsymbol{x} \in C_0)$ are distinct words of C_1 so $|C_0| \leq |C_1|$.

7. Let S be $\{\boldsymbol{x}_1, \boldsymbol{x}_2, \cdots, \boldsymbol{x}_k\}$. If $\boldsymbol{u}, \boldsymbol{v} \in \langle S\rangle$ and $\alpha, \beta \in Z_p$ we have $\alpha\boldsymbol{u}+\beta\boldsymbol{v} = \alpha\sum_{i=1}^{i} \alpha_i\boldsymbol{x}_i + \beta\sum_{i=1}^{k} \beta_i\boldsymbol{x}_i = \sum_{i=1}^{k}(\alpha\alpha_i + \beta\beta_i)\boldsymbol{x}_i \in \langle S\rangle$.

8. $(000013).(123142) = 0 + 0 + 0 + 0 + 4 + 6 \equiv 0 \bmod 5$, so 000013 is orthogonal to 123142.

9. If $\boldsymbol{u}$ is binary, of even weight, then $\boldsymbol{u}.\boldsymbol{u} = 1+1+\cdots+1$, where the number of 1s is just the number of 1s in $\boldsymbol{u}$, which is even, in other words, zero modulo 2.

If $\boldsymbol{u}$, $\boldsymbol{v}$ have the same weight w, let p, q, r, s be the number of positions in which $\boldsymbol{u}$, $\boldsymbol{v}$ have bits 0,0; 0,1; 1,0; 1,1 respectively. Then $w(\boldsymbol{u}) = r+s = w$ and $w(\boldsymbol{v}) = q+s = w$. From these, $d(\boldsymbol{u}, \boldsymbol{v}) = q+r = 2(w-s)$.

10.

$$x_1x_2x_3x_4 \in S^{\perp} \Leftrightarrow \left.\begin{array}{l} x_1 + 2x_2 + 2x_4 = 0 \\ x_1 + x_2 + x_3 + x_4 = 0 \\ 2x_1 = 0 \end{array}\right\}$$

$$\Leftrightarrow \left.\begin{array}{l} x_1 = 0, \\ x_2 + x_4 = 0 \\ x_2 + x_3 + x_4 = 0 \end{array}\right\}$$

$$\Leftrightarrow x_1 = 0, x_2 + x_4 = 0, x_3 = 0.$$

So $S^{\perp}$ is the set of words {0000, 0102, 0201}. Similarly $T^{\perp}$ is {00000, 00110, 11011, 11101}

11. $\boldsymbol{u}, \boldsymbol{v} \in S^{\perp} \Rightarrow \alpha\boldsymbol{u} + \beta\boldsymbol{v} \in S$, because for any $\boldsymbol{s} \in S$

$$(\alpha\boldsymbol{u} + \beta\boldsymbol{v}) \cdot \boldsymbol{s} = \alpha(\boldsymbol{u} \cdot \boldsymbol{s}) + \beta(\boldsymbol{v} \cdot \boldsymbol{s}) = \alpha 0 + \beta 0 = 0.$$

12. $\sum_{i=1}^{m} \alpha_i\boldsymbol{c}_i = \sum_{i=1}^{m} \beta_i\boldsymbol{c}_i \Rightarrow \sum_{i=1}^{m} (\alpha_i - \beta_i)\,\boldsymbol{c}_i = 0$

$\Rightarrow \alpha_i - \beta_i = 0$ for all i, by the definition of independence.

13. (i) $1111 = 1010 + 0101$ so $\langle 1010, 0101, 1111\rangle = \langle 1010, 0101\rangle$
$= \{\alpha(1010) + \beta(0101) : \alpha, \beta, \in Z_2\} = \{0000, 1010, 0101, 1111\}$

(ii) 0101, 1010, 1100 are independent so $\langle 0101,\ 1010,\ 1100\rangle$
$= \{\alpha(0101) + \beta(1010) + \gamma(1100)\ :\ \alpha, \beta, \gamma, \in\ Z_2\}$
$= \{0000,\ 0101,\ 1010,\ 1100,\ 1111,\ 1001,\ 0110,\ 0011\}$.

(iii) 10101 is the sum of the other three so the span is
$\langle 00111, 01011, 11001\rangle\ =\ \{00000, 00111, 01011, 11001, 01100, 10010, 11110, 10101\}$

(iv) $\{0000,\ 1011,\ 2022,\ 0112,\ 0221,\ 1120,\ 1202,\ 2101,\ 2210\}$

14. This can be done by throwing out members of these sets which are linear combinations of the other members. A better method will be given later.

(i) $\{1010,\ 1001,\ 0101\}$

(ii) $1000 = 2(3410) + 2(0140)$,

so throw out 1000. Then $3410\ =\ 1234\ +\ 3(0140)\ +\ 3(4322)$, so throw out 3410. The remaining set $\{0140,\ 4322,\ 1234\}$ is independent, so is a basis of the span of the original set. To extend this to a basis of Z_5^4 we need to add one more word so that the resulting set is independent. The general member of $\langle 0140,\ 4322,\ 1234\rangle$ is $\alpha(0140) + \beta(4322) + \gamma(1234) = 4\beta + \gamma, \alpha + 3\beta + 2\gamma,\ 4\alpha + 2\beta + 3\gamma,\ 2\beta + 4\gamma$.

Now any word of this form has 0 for the sum of its 2nd and 3rd symbols. A simple word not having this property is 0010, so

$$\langle 0140,\ 4322,\ 1234,\ 0010\rangle\ = Z_5^4$$

and we have our basis.

16. Let $\{\boldsymbol{c}_1,\ \boldsymbol{c}_2,\ \cdots\ \boldsymbol{c}_k\}$ be a basis for the code. The codewords are all the words of the form $\sum_{i=1}^{k} \alpha_i\ \boldsymbol{c}_i$, and since each α_i may be freely chosen from the p symbols of Z_p, there are p^k different choices for the sequence of coefficients $\alpha_1, \cdots,\ \alpha_k$. By Exercise 12 these correspond to distinct codewords, so the size of the code is p^k.

17. (i) Let B_1 be any basis of C_1. Then B_1 is a linearly independent set of words of C_2. Hence it can be extended to a basis of C_2, so B_1 must have at most k_2 words.

(ii) If $k_1 = k_2$ then B_1 above is already a basis for C_2, so $C_1 = C_2$ since they are both equal to $\langle B_1\rangle$.

18. Suppose $\boldsymbol{m}_1, \boldsymbol{m}_2$ are distinct messages so that $\boldsymbol{m}_1 - \boldsymbol{m}_2 \neq \boldsymbol{0}$. Then if these messages correspond to the same codeword, $\boldsymbol{m}_1 G = \boldsymbol{m}_2 G \Rightarrow (\boldsymbol{m}_1 - \boldsymbol{m}_2)G = \boldsymbol{0}$
$\Rightarrow$ the k rows of G are independent which is false (see definition of a generator matrix and of a basis).

19. 02204 (Add rows 1 and 3).

21. (i) $C+\boldsymbol{x}$ is formed by adding $\boldsymbol{x}$ to each $\mathbf{c} \in C$, so we get a list of $|C|$ words. To show that the coset size is $|C|$ we just need to show that no two words in the list are the same. This is clear because if $\mathbf{c}, \mathbf{c}'$ are distinct words of C then $\mathbf{c}+\boldsymbol{x} \neq \mathbf{c}'+\boldsymbol{x}$.

(ii) $\boldsymbol{y} \in C+\boldsymbol{x} \Rightarrow \boldsymbol{y} = \mathbf{c}+\boldsymbol{x}$ for some $\mathbf{c} \in C$. Let $\boldsymbol{z} \in C+\boldsymbol{y}$, so $\boldsymbol{z} = \mathbf{c}'+\boldsymbol{y}$ for some $\mathbf{c}' \in C$, so $\boldsymbol{z} = \mathbf{c}'+\mathbf{c}+\boldsymbol{x} \in C+\boldsymbol{x}$. Therefore $C+\boldsymbol{y} \subseteq C+\boldsymbol{x}$ and a similar argument proves $C+\boldsymbol{x} \subseteq C+\boldsymbol{y}$. To prove the second part, suppose $\exists \boldsymbol{z}$ such that $\boldsymbol{z} \in (C+\boldsymbol{x}) \cap (C+\boldsymbol{y})$. Then $\boldsymbol{z} = \mathbf{c}+\boldsymbol{x} = \mathbf{c}'+\boldsymbol{y}$ for some $\mathbf{c}$, $\mathbf{c}'$ in C. This implies $y = \mathbf{c}-\mathbf{c}'+\boldsymbol{x} \in C+\boldsymbol{x}$, a contradiction.

(iii) Let $\boldsymbol{x}$ be any word of Z_p^n. Then $\boldsymbol{x} = \boldsymbol{x}+\mathbf{0}$ and $\mathbf{0} \in C$ so $\boldsymbol{x} \in \boldsymbol{x}+C$.

(iv) $\boldsymbol{x}, \boldsymbol{y} \in C+\boldsymbol{z} \Rightarrow \boldsymbol{x} = \mathbf{c}+\boldsymbol{z}, \boldsymbol{y} = \mathbf{c}'+\boldsymbol{z}$ for some $\mathbf{c}, \mathbf{c}' \in C$

$$\Rightarrow \boldsymbol{x}-\boldsymbol{y} = \mathbf{c}-\mathbf{c}' \in C$$

and $\boldsymbol{x}-\boldsymbol{y} \in C \Rightarrow \boldsymbol{x}-\boldsymbol{y} = \mathbf{c}$ for some $\mathbf{c} \in C \Rightarrow \boldsymbol{x} = \mathbf{c}+\boldsymbol{y}$

$$\Rightarrow \boldsymbol{x} \in C+\boldsymbol{y} \text{ and } \boldsymbol{y} \in C+\boldsymbol{y} \text{ (see proof of (iii))}$$

(v) From (ii) any two cosets $C+\boldsymbol{x}, C+\boldsymbol{y}$ are either disjoint or identical. From (i) each coset has size $|C|$, and from (iii) their union is Z_p^n, which has size p^n. So the number of distinct cosets is $p^n/|C| = p^n/p^k = p^{n-k}$.

22. C :

00000	10011	01011	00101	01110	10110	11000	11101
10000	00011	11011	10101	11110	00110	01000	01101
00100	10111	01111	00001	01010	10010	11100	11001
00010	10001	01001	00111	01100	10100	11010	11111

In this case there is some freedom of choice for the Slepian array. 00000 and 00010 must be coset leaders, but for the other two one can have either 10000 or 01000 as leader and the other can have 00100 or 00001. But $\boldsymbol{r} = 01100$ occurs in a coset for which there is no choice of leader, so must be decoded as 01110.

$$C' = \{0000,\ 0121,\ 1220\ 0212,\ 2110,\ 1011,\ 2201,\ 1102,\ 2022\}$$

$$|C'| = 9 \text{ and } |Z_3^4| = 81$$

so there are $\frac{81}{9} = 9$ cosets.

23. In the array for C 00001 is not a coset leader, so this is an error pattern of weight 1 which is not correctly decoded. For example, if $\mathbf{c} = 00000$ and $\boldsymbol{r} = 00001$, then $\boldsymbol{r}$ is decoded as 00101. C' has minimum non-zero weight 3, so this is its minimum distance. Hence it is 1-error correcting, so every word of weight 1 must be a coset leader in any Slepian array. There are 8 words of weight 1 and 9 cosets, the first of which is C' itself - having $\mathbf{0}$ as its leader. So 0 and 1 are the only coset leader weights.

24. By Theorem 4.10 the sum of any two even weight words is another even weight word. Also, if $\boldsymbol{x}$ has even weight, so do $1\boldsymbol{x}$ and $0\boldsymbol{x}$, so C is linear. Let $\boldsymbol{w}$ be any word with odd weight. By Theorem 4.10 again, the coset

$C + \boldsymbol{w}$ consists entirely of odd weight words. Furthermore this coset contains all the words of odd weight because any odd weight word $\boldsymbol{v}$ can be written as $(\boldsymbol{v} + \boldsymbol{w}) + \boldsymbol{w} \in C + \boldsymbol{w}$ since $\boldsymbol{v} + \boldsymbol{w}$ has even weight. Hence C consists of half the words of Z_2^n, so $\dim(C) = n - 1$. So if G generates C, G is an $(n-1) \times n$ matrix, and to be in standard form its first $n - 1$ columns make up I_{n-1}. Every row of G is a codeword, so has even weight, so this forces the last column to be all 1s.

$C^{\perp}$ has dimension 1 so its generator matrix is a single row. If $\mathbf{c}' = x_1x_2 \cdots x_n$ is any word of $C^{\perp}$, it must be orthogonal to every word in C, in particular to every row of G. This leads to the equations $x_i + x_n = 0 (i = 1, 2, \cdots, n-1)$, the only solutions of which are $x_i = 0$ for all i or $x_i = 1$ for all i. So the only generator for $C^{\perp}$ is $[11 \cdots 1]$ (and this *is* in standard form).

25. Let X, Y be the sets of even and odd weight words respectively of C. Suppose $Y \neq \phi$ so $\exists\, \boldsymbol{y} \in Y$. Show that the members of $\boldsymbol{y} + X$ are all different and belong to Y and that the members of $\boldsymbol{y} + Y$ are all different and belong to X (using an obvious 'coset notation' even though Y is not a linear code). Hence $|Y| \geq |X|$ and $|X| \geq |Y|$ so $|X| = |Y|$.

    If C has a generator matrix in which all rows have even weight, then every codeword has even weight by applying Theorem 4.10.

26. First note that since the rows of G_1 are independent, so are the rows of G (why?), so G is a genuine generator matrix. Let $\boldsymbol{r}_i$, $\boldsymbol{s}_i (i = 1, 2, \cdots k)$ be the rows of G_1, G_2 respectively. In the notation following Theorem 4.12 any word of C can be written

$$\sum_{i=1}^{k} \lambda_i \left(\boldsymbol{r}_i \middle| \boldsymbol{s}_i\right) = \left(\sum_{i=1}^{k} \lambda_i\, \boldsymbol{r}_i\right) \middle| \left(\sum_{i=1}^{k} \lambda_i \boldsymbol{s}_i\right).$$

    Provided not all the λ_i are zero, this is of the form $\mathbf{c}_1|\mathbf{c}_2$ where $\mathbf{c}_1$, $\mathbf{c}_2$ are non-zero words of C_1, C_2 respectively. Now $w(\mathbf{c}_1|\mathbf{c}_2) = w(\mathbf{c}_1) + w(\mathbf{c}_2)$ and $w(\mathbf{c}_1) \geq d_1$ and $w(\mathbf{c}_2) \geq d_2$.

    Hence all non-zero words of C have weight $\geq d_1 + d_2$ so $d(C) \geq d_1 + d_2$.

27. Let $\mathbf{c}$, $\mathbf{c}' \in C \cup (C + \boldsymbol{a})$. Show $\mathbf{c} + \mathbf{c}' \in C \cup (C + \boldsymbol{a})$ by checking the three cases : $\mathbf{c}$, $\mathbf{c}'$ both in C, both in $C + \boldsymbol{a}$, or one in each.

28. Write the Venn diagram conditions as equations modulo 2 as in section 5.6 to get the 8-bit code described as $S^{\perp}$ where

$$S = \{10111000,\ 11010100,\ 11100010,\ 11111111\}.$$

29. Let $\boldsymbol{x} \in C$. The $\boldsymbol{x} \cdot \boldsymbol{y} = 0$ for all $\boldsymbol{y} \in C^{\perp}$, so $\boldsymbol{x} \in (C^{\perp})^{\perp}$. Hence $C \subseteq (C^{\perp})^{\perp}$. But $\dim C = \dim((C^{\perp})^{\perp})$ by Theorem 5.6.4, so by applying (11) (ii) of section 5.3 $C = (C^{\perp})^{\perp}$.

30. The repetition code C of length n over Z_p is a code in which there are p codewords; $\boldsymbol{i} = i\ i\ i\ \cdots i,\ i = 0, 1, 2, \cdots, p-1$. Such a code is clearly linear. Let $\boldsymbol{x} = x_1 x_2 \cdots x_n$ be any word. Then $\boldsymbol{x} \cdot \boldsymbol{i} = 0$ if and only if $i(x_1 + x_2 + \cdots + x_n) = 0$. So $\boldsymbol{x}$ is orthogonal to every word of C if and only if $x_1 + x_2 + \cdots + x_n = 0$. That is $C^\perp = S$ where $S = \{\boldsymbol{x} : \boldsymbol{x} \in Z_p^n,\ x_1 + \cdots + x_n = 0\}$, so $(C^\perp)^\perp = C = S^\perp$.

31. G is a 2×4 matrix, so dim $C = 2$ and hence dim $(C^\perp) = 4 - 2 = 2$. So the parity check matrix will also be 2×4. To find one we have to find two independent rows each of which is orthogonal to every word of C. To ensure this it is enough to have the two rows orthogonal to each row of G (why?), so the equations to be satisfied (modulo 3) are $x_1 + x_2 + x_3 = 0,\ 2x_1 + x_3 + x_4 = 0$, for which one (of many) independent pair of solutions is 1110 and 0012, so

$$H = \begin{bmatrix} 1 & 1 & 1 & 0 \\ 0 & 0 & 1 & 2 \end{bmatrix}$$

is a suitable parity check matrix.

32. The matrix has all its rows of weight at least 3 so $d(C) \leq 3$. The sub-words to the right of I_7 are distinct, so the sum of any two rows has weight ≥ 3 (2 from the I_7 entries and at least 1 from positions 8-11). Finally, just from the I_7 entries it is clear that the sum of any three or more rows has weight at least 3. Hence $d(C) = 3$.

33. Let the ith column of H be $\mathbf{0}$, and let $\mathbf{c}$ be a non-zero codeword received as $\boldsymbol{r}$ with an error in bit i and with no other error. Then by Theorem 5.11 syn$(\boldsymbol{r}) = \mathbf{0}$ so $\boldsymbol{r}$ will be decoded (incorrectly) to $\boldsymbol{r}$, which contradicts the fact that C is 1-error correcting.

    Now suppose columns i and j are the same and let $\mathbf{c}, \mathbf{c}'$ be any two distinct codewords. Let $\boldsymbol{r},\ \boldsymbol{r}'$ be the words obtained by changing bits $i,\ j$ respectively of $\mathbf{c}$. Then $\boldsymbol{r}$ and $\boldsymbol{r}'$ have the same syndrome, so are decoded to the same codeword. Hence at least one decoding is incorrect, again contradicting C being 1-error correcting.

34. The syndrome equations are

$$\begin{array}{ccccccccccccccc} a & + & 2b & & & + & d & & & & & & & = & 1 \\ a & + & b & + & c & & & + & e & & & & & = & 1 \\ & & & & c & & & & & + & f & & & = & 1 \\ 2a & + & 2b & + & 2c & & & & & & & + & g & = & 1 \end{array}$$

From the third equation there are just three possible values of (c, f): (0, 1); (1, 0); (2, 2). Taking the first of these, the syndrome equations reduce to $a + 2b + d = 1;\ a + b + e = 1;\ 2a + 2b + g = 1$ and we require a solution in which at most one of a, b, d, e and g is non-zero. Taking all except a to be zero the equations become $a = 1,\ a = 1$ and $2a = 1$, which are clearly incompatible. Similar arguments rule out all other possibilities.

This means that all words whose syndrome is 1111 must have a weight of at least 3. So the code will correct all received words with one error, not all words with two errors, and will correct some with more than two errors.

35. The matrix will be 3×6 with no set of three dependent columns. There are only 7 non-zero columns so H must have all but one of these. Hence it must have all the columns of weight 2, or all the columns of weight 1 together with at least two of weight 2. In both cases it is clear that three columns will be dependent.

36.(a) 3 (columns 1, 6, 11)

(b) 4 (columns 3, 4, 7, 8). Since all columns are distinct and have odd weight, no three of them can have zero sum.

(c) 2 (columns 3, 7)

(d) 1 (column 5).

37. Since the code is perfect with $d = 3$ the coset leaders are all the words of length 7 with weights 0 or 1 (8 leaders in all).
G is already in standard form so

$$H = \begin{bmatrix} 1 & 1 & 1 & 0 & 1 & 0 & 0 \\ 1 & 1 & 0 & 1 & 0 & 1 & 0 \\ 1 & 0 & 1 & 1 & 0 & 0 & 1 \end{bmatrix}$$

and the coset leader with 1 in the ith place is just the ith column of H.

syn(0000011) = 011, so error is in the 4th place
syn(1111111) = 000, so there is no error.
syn(1100110) = 111, so error is in the 1st place.

Hence the decoded words are 0001011, 1111111 and 0100110.

38. The generator matrix has $\binom{1}{0}$ and $\binom{0}{1}$ as two of its columns, so the algorithm gives the parity check matrix,

$$H = \begin{bmatrix} 1 & -1 & 0 & -2 \\ 0 & -1 & 1 & -1 \end{bmatrix} = \begin{bmatrix} 1 & 2 & 0 & 1 \\ 0 & 2 & 1 & 2 \end{bmatrix}$$

This code has distance 3, so all words of weights 0 and 1 are coset leaders. There are only 9 coset leaders in all so this accounts for all of them.

syn(2121) = 20 = 2 × col 1 $\Rightarrow \mathbf{e} = 2000$ so decoded word is 0121
syn(1201) = 00 $\Rightarrow \mathbf{e} = 0000$ so decoded word is 1201
syn(2222) = 21 = 2 × col 4 $\Rightarrow \mathbf{e} = 0002$ so decoded word is 2220

39. $G' = \begin{bmatrix} 2 & 0 & 1 \\ 1 & 1 & 0 \end{bmatrix}$ so the two codes are

C = 000 201 102 112 221 010 122 211 020
C' = 000 201 102 110 220 011 121 212 022

These are clearly not equivalent as $d(C) = 1, d(C') = 2$. Also in position 3, 1 is sometimes fixed (as in the second word) and sometimes not.

40. $G' = \begin{bmatrix} 2 & 2 & 1 \\ 1 & 1 & 2 \end{bmatrix}$

and this is not a generator matrix since its rows are dependent.

41. Let

$$G = \begin{bmatrix} 2 & 1 & 1 \\ 1 & 0 & 2 \end{bmatrix}$$

generate C over Z_3, and let

$$G' = \begin{bmatrix} 1 & 2 & 1 \\ 0 & 1 & 2 \end{bmatrix},$$

$$G'' = \begin{bmatrix} 1 & 1 & 1 \\ 2 & 0 & 2 \end{bmatrix}$$

generate C', C'' respectively. [For G' interchange columns 1 and 2 of G; for G'' multiply column 1 of G by 2.] Then 211 is a codeword of C but not of C' or C''.

42. Clearly R1 only changes the order in which the basis words of C are written down. For a representative example of R2 let G have rows $\boldsymbol{r}_i (i = 1, 2, \cdots, k)$ and let G' have rows $\boldsymbol{r}'_i$ where $\boldsymbol{r}'_1 = a\boldsymbol{r}_1 (a \neq 0)$, and for all $i \geq 2, \boldsymbol{r}'_i = \boldsymbol{r}_i$.

$$\begin{aligned}
\text{Then } \boldsymbol{c} \in C \quad \Leftrightarrow \quad \boldsymbol{c} \quad &\in \quad \langle \boldsymbol{r}_1, \boldsymbol{r}_2, \cdots, \boldsymbol{r}_k \rangle \\
\Rightarrow \quad \boldsymbol{c} \quad &= \quad \sum_{i=1}^{k} \lambda_i \boldsymbol{r}_i \Rightarrow \boldsymbol{c} = \lambda_1 a^{-1}(a\boldsymbol{r}_1) + \sum_{i=2}^{k} \lambda_i \boldsymbol{r}_i \\
\Rightarrow \quad \boldsymbol{c} \quad &= \quad \lambda_1 a^{-1}(\boldsymbol{r}'_1) + \sum_{i=2}^{k} \lambda_i \boldsymbol{r}'_i \\
\Rightarrow \quad \boldsymbol{c} \quad &\in \quad \langle \boldsymbol{r}'_1, \boldsymbol{r}'_2, \cdots, \boldsymbol{r}'_k \rangle = C'.
\end{aligned}$$

The reverse implication is shown in a similar way, and of course the argument is identical if some row other than the first is changed.

Finally, for R3, let $\boldsymbol{r}'_1 = \boldsymbol{r}_1 + a\boldsymbol{r}_2$ and the other rows are unchanged.

$$\begin{aligned}
\text{Then } \boldsymbol{c} \in C \Rightarrow \boldsymbol{c} = \sum_{i=1}^{k} \lambda_i \boldsymbol{r}_i \Rightarrow \boldsymbol{c} \quad &= \quad \lambda_1(\boldsymbol{r}'_1 - a\boldsymbol{r}_2) + \sum_{i=2}^{k} \lambda_i \boldsymbol{r}_i \\
&= \quad \lambda_1 \boldsymbol{r}'_1 + (\lambda_2 - a\lambda_1)\boldsymbol{r}'_2 + \sum_{i=3}^{k} \lambda_i \boldsymbol{r}'_i \\
&\in \quad \langle \boldsymbol{r}'_1, \boldsymbol{r}'_2, \cdots, \boldsymbol{r}'_k \rangle = C'.
\end{aligned}$$

$$\text{and } \mathbf{c}' \in C' \Rightarrow \mathbf{c}' = \sum_{i=1}^{k} \mu_i \boldsymbol{r}_i' \Rightarrow \mathbf{c}' \;=\; \mu_1(\boldsymbol{r}_1 + a\boldsymbol{r}_2) + \sum_{i=2}^{k} \mu_i \boldsymbol{r}_i$$

$$= \mu_1 \boldsymbol{r}_1 + (a\mu_1 + \mu_2)\boldsymbol{r}_2 + \sum_{i=3}^{k} \mu_i \boldsymbol{r}_i$$

$$\in \langle \boldsymbol{r}_1, \boldsymbol{r}_2, \cdots, \boldsymbol{r}_k \rangle = C.$$

43. No solution given because the final form depends on which non-zero entries you select for each application of stage (a).

44. $syn(1xyz12) = (x + 2y + z,\ x + 2y + z,\ x + 2z) = (0,\ 0,\ 0)$ if and only if $(x, y, z) = (0,0,0), (1,2,1)$ or $(2,1,2)$.
$syn(xyz210) = (2x + y + 2z,\ x + y + 2z + 2,\ y + 1) = (0,\ 0,\ 0)$ if and only if $(x, y, z) = (2, 2, 0)$
So only the second word is uniquely decodable.

45. Such a code has a 4×10 parity check matrix, so its columns are length 4. Any 5 vectors of length 4 must be dependent.

46. For both codes $d = 2$ so both are 1-erasure decodable, but not 2-erasure decodable. However the first has three pairs of dependent columns (24, 25 and 45) but the second has only two (15 and 23). Hence the second code will decode uniquely more instances of double erasures than will the first code.

47. There are four possible codewords consistent with a single error, and there must be at least one error since the received word has syndrome $(3x,\ x + 2)$. They are $\mathbf{c} = 04421, 14223, 22423$ and 34433.

48. $syn(\boldsymbol{r}) = (1, 0, 1+x)$ which is never $\mathbf{0}$. Hence there is an error. $syn(\boldsymbol{r}) = 101, 100$ if $x = 0, 1$ respectively. Only 100 is a column of H, so $\mathbf{c} = 101101 - 000100 = 101001$.
From the given H, G can be found:

$$G = \begin{bmatrix} 1 & 0 & 0 & 1 & 1 & 0 \\ 0 & 1 & 0 & 0 & 1 & 1 \\ 0 & 0 & 1 & 1 & 1 & 1 \end{bmatrix},$$

then

$$C = \{000000, 100110, 010011, 001111, 110101, 101001, 011100, 111010\}.$$

For $x = 0, \boldsymbol{r} = 101100$, and for $x = 1$ $\boldsymbol{r} = 101101$. Neither are codewords, the first is at distance at least 2 from all codewords, the second has distance 1 from codeword 101001 and a greater distance from all others.

49. $\boldsymbol{r} = 1x20y1$. In this case the received word is a codeword (has syndrome $\mathbf{0}$) if and only if $x = y = 0$, so we decode to 102001.

$\boldsymbol{r} = 21xy11$. In this case the syndrome is $(2x+y,\ 2x+y+1,\ 2y+1)$ which is never $\mathbf{0}$, so there is at least one error. Checking the nine possibilities for $(x,\ y)$, only the following three cases give a syndrome which is a multiple of a column of H.

These are:

xy	syndrome		
01	120	=	$2\times$ column 1
11	010	=	$1\times$ column 6
21	200	=	$2\times$ column 5,

so the three most likely transmitted codewords (those involving only one error) are 010111, 211110, 212121.

Chapter 6

1. The partition of the non-zero columns is

$$\left\{\begin{matrix} 0 & 0 & 0 & 0 \\ 1 & 2 & 3 & 4 \end{matrix}\right\},\ \left\{\begin{matrix} 1 & 2 & 3 & 4 \\ 2 & 4 & 1 & 3 \end{matrix}\right\},\ \left\{\begin{matrix} 2 & 4 & 1 & 3 \\ 3 & 1 & 4 & 2 \end{matrix}\right\},$$
$$\left\{\begin{matrix} 3 & 1 & 4 & 2 \\ 4 & 3 & 2 & 1 \end{matrix}\right\},\ \left\{\begin{matrix} 4 & 3 & 2 & 1 \\ 0 & 0 & 0 & 0 \end{matrix}\right\},\ \left\{\begin{matrix} 1 & 2 & 3 & 4 \\ 1 & 2 & 3 & 4 \end{matrix}\right\},$$

so a possible H is $\begin{bmatrix} 0 & 1 & 2 & 3 & 4 & 1 \\ 1 & 2 & 3 & 4 & 0 & 1 \end{bmatrix}$.

2. Let $\boldsymbol{h}, \boldsymbol{h}'(\boldsymbol{h} \neq \boldsymbol{h}')$ be columns selected as suggested, and let their first non-zero symbols be the ith and jth respectively. Suppose $\boldsymbol{h} = \alpha\boldsymbol{h}'$ for some non-zero α. Then if $i = j$, α can only be 1, which contradicts $\boldsymbol{h} \neq \boldsymbol{h}'$. And if $i \neq j$ (say $i < j$), then $\boldsymbol{h} = \alpha\boldsymbol{h}'$ is impossible because $\boldsymbol{h}'$ has 0 in its ith place but $\boldsymbol{h}$ has $\alpha \neq 0$ in its ith place.

   We also get the right (maximal) number of columns in this way : there are q^{r-i} columns whose first non-zero symbol is the ith (the first $i-1$ are 0, the ith is 1, and there is a free choice of any of the q symbols in each of the remaining $r-i$ places). So the total number of columns is

$$\sum_{i=1}^{r} q^{r-i} = \frac{q^r - 1}{q-1}\ ,\ \text{as required by Theorem 6.1}$$

3. Clearly no pair of these three columns are dependent so they belong to three distinct subsets $m(\boldsymbol{u})$, $m(\boldsymbol{v})$, $m(\boldsymbol{w})$. The columns of H must include a representative from each of these. That is, it contains columns of the given forms.

4. $syn(\boldsymbol{r})$ = sum of first 4 columns of $H = (00100)^T$, which represents 4 in binary. Hence the decoded word is $11110\cdots0$.

5. $n = 8,\ k = 6$, so $dim(C^{\perp}) = 8 - 6 = 2 = r$. Hence $n = \dfrac{q^2-1}{q-1} = q+1$, so $q = 7$.

   So $H = \begin{bmatrix} 0 & 1 & 1 & 1 & 1 & 1 & 1 & 1 \\ 1 & 0 & 1 & 2 & 3 & 4 & 5 & 6 \end{bmatrix}$.

syn(12312300) = (4, 3) = 4(1, 6) = 4 × column 8

so decoded word = 12312300 − 00000004 = 12312303.

6. Reduce G to nearly standard form by row operations, and hence obtain H, which contains all the non-zero columns.

7. The proof of Theorem 6.6 showed that all non-zero codewords of $C^\perp$ have weight q^{r-1}, so $d(C^\perp) = q^{r-1}$. Hence $C^\perp$ is t-error correcting, where

$$t = \left\lfloor \frac{q^{r-1} - 1}{2} \right\rfloor$$

8. Let $d(C) = w$, so each of the 80 non-zero rows of the array contains w non-zero symbols. Let x be the number of all zero columns in the array. By Theorem 5.4 the remaining $10 - x$ columns have $\frac{2}{3} \times 81 = 54$ non-zeros each. Hence $80w = 54(10 - x)$, which implies $40w \equiv 0 \mod 27$, and since $gcd(40, 27) = 1$, $w \equiv 0 \mod 27$ is the only solution. But since C is 2-error correcting and has length 10, $5 \leq w \leq 10$, so there is no solution in this range.

9. A binary simplex code is $C^\perp$ for some $C \in \text{Ham}(r, 2)$. $C^\perp$ has 2^r codewords of length $2^r - 1$, and $d(C^\perp) = 2^{r-1}$ from the proof of Theorem 6.6. From these it follows that $C^\perp$ meets the Plotkin bound and is therefore optimal.

10. Without taking linearity into account we have $6 \leq M \leq 51$ from the G–V and Hamming bounds. (The Hamming bound is stronger than the Singleton, and the Plotkin bound does not apply for these parameters.) Since M must be a power of 2 for a linear code this estimate can be improved to $M = 8$, 16 or 32.

11. Use induction to show that for all $i \leq k$, the set $\{\mathbf{c}_1, \cdots, \mathbf{c}_i\}$ is independent.

12. Let C be a binary (n, M, n) code. As usual we can change this to an equivalent code C' in which one word is $\mathbf{0}$. Since $d = n$, every symbol in all other words is non-zero, so there is only one other word, which has to be $111\cdots1$.

13. A binary repetition code of even lenth n has $d = n$, and by Exercise 15 of Chapter 4, all perfect codes have odd minimum distance. A repetition non-binary code with alphabet size $q \geq 3$ has $M = q$ and $d = n$. It is therefore $(\frac{n-1}{2})$-error correcting. We know n is odd if the code is to be perfect. For $C = \{000, 111, 222\}$ check that C does not meet the Hamming bound. For $n \geq 5$ consider sending $\mathbf{0}$ and receiving $\boldsymbol{r}$ which has $\frac{n-1}{2}$ zero symbols, and $\frac{n+1}{2}$ non-zeros made up of 2 αs and $\frac{n-3}{2}$ βs ($\alpha \neq \beta$). The $\mathbf{0}$ is the nearest codeword to $\boldsymbol{r}$ so $\boldsymbol{r}$ is correctly decoded even though it contains more than $\frac{d-1}{2}$ errors. This contradicts Exercise 16 of Chapter 4.

14. Let

$$G = \left[\begin{array}{cccc|cccc} 0 & 0 & \cdots & 0 & 1 & 1 & \cdots & 1 \\ & & & & & & & \\ & & G_1 & & & & G_2 & \end{array}\right],$$

$$G' = \left[\begin{array}{cccc|cccc} 0 & 0 & \cdots & 0 & 1 & 1 & \cdots & 1 \\ & & & & & & & \\ & & G'_1 & & & & G'_2 & \end{array}\right]$$

be generator matrices for C with equal first rows. We have to show that the span of the rows of G_1 is the same as the span of the rows of G'_1.

Let $\boldsymbol{u}_1 \in \langle \text{rows of } G_1 \rangle$ and let $\boldsymbol{u}_1|\boldsymbol{u}_2$ be the corresponding combination of rows of $[G_1|G_2]$. Then $\boldsymbol{u}_1|\boldsymbol{u}_2 \in C$ so is in $\langle$ rows of $G' \rangle$. Hence $\boldsymbol{u}_1$ is in the span of the left hand block of G', and since the first word of this block is $\mathbf{0}$, $\boldsymbol{u}_1 \in \langle$ rows of $G'_1 \rangle$. Hence $\langle$ rows of $G_1 \rangle \subseteq \langle$ rows of $G'_1 \rangle$ and the proof of the reverse inclusion is identical.

15. Let C have generator matrices

$$G = \begin{bmatrix} 1 & 1 & 1 & 1 & 1 & 0 & 0 \\ 0 & 1 & 0 & 1 & 1 & 1 & 1 \\ 0 & 0 & 1 & 1 & 0 & 0 & 0 \end{bmatrix}$$

and

$$G' = \begin{bmatrix} 0 & 1 & 0 & 1 & 1 & 1 & 1 \\ 1 & 1 & 1 & 1 & 1 & 0 & 0 \\ 0 & 0 & 1 & 1 & 0 & 0 & 0 \end{bmatrix}$$

where rows 1 and 2 (both of weight 5) have been switched to obtain G' from G. Then

$$G_1 = \begin{bmatrix} 1 & 1 \\ 0 & 0 \end{bmatrix}$$

and

$$G'_1 = \begin{bmatrix} 1 & 1 \\ 0 & 1 \end{bmatrix},$$

so the two residual codes have different dimensions.

16. The codes have dimension at least 2, so have at least 4 codewords, so must include words other than $\mathbf{0}$ and the all 1s word.

17. x can be written as $2a+r$ where a is an integer and $0 \leq r < 2$. Check the claimed result by checking each of the cases: $r = 0$; $0 < r < 1$; $r = 1$; $1 < r < 2$.

18. Use induction. The case $i = 1$ is just Theorem 6.11. Assuming the

result for i, the inductive step is:

$$\begin{aligned} d(Res^{i+1}C) &= d(Res(Res^iC)) \\ &\geq \left\lceil \frac{d(Res^iC)}{2} \right\rceil && \text{by the theorem} \\ &\geq \left\lceil \frac{\left\lceil \frac{d}{2^i} \right\rceil}{2} \right\rceil && \text{by the induction hypothesis} \\ &= \left\lceil \frac{d}{2^{i+1}} \right\rceil && \text{by Exercise 17.} \end{aligned}$$

19. The Hamming, Plotkin and Griesmer bounds give $k \leq 4, 3, 2$ respectively.

20. Let B_k be the binary matrix whose columns are all the binary strings of length k. If the all zero column is deleted this leaves the parity check matrix for Ham$(k, 2)$ – that is, the generator matrxi G_k of the binary simplex code of dimension k. By symmetry each row of B_k will contain 2^{k-1} zeros and 2^{k-1} ones, and therefore has weight 2^{k-1} which is even provided $k \geq 2$. Similarly in any two distinct pairs of rows of B_k the bit pairs
$$\begin{matrix} 0 & 0 & 1 & 1 \\ 0, & 1, & 0, & 1 \end{matrix}$$
each occur 2^{k-2} times so the dot product of these rows is 0 provided $k \geq 3$. These conclusions hold for the rows of G_k too since deleting the all zero column makes no difference to the even weight property nor to the orthogonality of the pairs of rows. Hence G_k satisfies the conditions of Theorem 6.14.

21. The problem is that $w(\boldsymbol{r})$ being a multiple of p does not guarantee that $\boldsymbol{r} \cdot \boldsymbol{r} = 0$. For example, over Z_5, $w(11112) = 5$ but $\boldsymbol{r} \cdot \boldsymbol{r} = 3$.

22. $\begin{bmatrix} 1 & 1 & 1 & 1 & 1 \\ 1 & 2 & 3 & 4 & 0 \end{bmatrix}$ and $\begin{bmatrix} 1 & 1 & 1 & 1 & 2 \\ 1 & 2 & 3 & 4 & 0 \end{bmatrix}$ are generator matrices which clearly generate equivalent codes over Z_5, but only the first code is self-orthogonal.

23. For $C \in$ Ham$(r,\ q)$ the necessary condition that the dimension is half the length becomes
$$r = \frac{1}{2} \cdot \frac{q^r - 1}{q - 1}.$$
$r = 2$, $q = 3$ is the only solution, and the code in Ham(2, 3) with generator matrix
$$\begin{bmatrix} 1 & 1 & 1 & 0 \\ 0 & 1 & 2 & 1 \end{bmatrix}$$
is self-dual.

24.

$$G = \begin{bmatrix} 1 & 1 & 0 & 0 & 0 & 0 & 0 & 0 & 0 & 0 \\ 0 & 0 & 1 & 1 & 0 & 0 & 0 & 0 & 0 & 0 \\ 0 & 0 & 0 & 0 & 1 & 1 & 0 & 0 & 0 & 0 \\ 0 & 0 & 0 & 0 & 0 & 0 & 1 & 1 & 0 & 0 \\ 0 & 0 & 0 & 0 & 0 & 0 & 0 & 0 & 1 & 1 \end{bmatrix}$$

is an easy to check though not very 'good' example.

25. Every word of C is $\mathbf{0}$ or a sum of rows of G. Clearly $w(\mathbf{0})$ is a multiple of 4. Let the codeword $\boldsymbol{c}(\neq \mathbf{0})$ be the sum of m rows of G, and use induction on m. We are given that the claimed result is true for $m = 1$, so assume it holds for all codewords with $m = l$. Let $\boldsymbol{c}'$ be a sum of $l+1$ rows of G, so that $\boldsymbol{c}'$ can be written as $\boldsymbol{x}+\boldsymbol{y}$ where $\boldsymbol{x}$ is a sum of l rows and $\boldsymbol{y}$ is a single row. Then $w(\boldsymbol{c}') = w(\boldsymbol{x}) + w(y) - 2w(\boldsymbol{x} \odot \boldsymbol{y})$, and the first two terms on the right hand side are multiples of 4. To deal with the remaining term, $w(\boldsymbol{x} \odot \boldsymbol{y})$ is even if $\boldsymbol{x}$, $\boldsymbol{y}$ are orthogonal, and odd otherwise, but $\boldsymbol{x}$, $\boldsymbol{y}$ *are* orthogonal since they are codewords of a self-orthogonal code. Hence $2w(\boldsymbol{x} \odot \boldsymbol{y})$ is a multiple of 4.

27. Codes C' and D' both have $16 = 2^4$ codewords. Hence there are $(2^4)^3 = 2^{12}$ choices of the triple $(\boldsymbol{a}, \boldsymbol{b}, \boldsymbol{x})$. We show that they correspond to 2^{12} distinct codewords of E', or equivalently, if $(\boldsymbol{a}', \boldsymbol{b}', \boldsymbol{x}')$ is also a triple from $C' \times C' \times D'$, the E-words $\boldsymbol{a}+\boldsymbol{x}|\boldsymbol{b}+\boldsymbol{x}|\boldsymbol{a}+\boldsymbol{b}+\boldsymbol{x}$ and $\boldsymbol{a}'+\boldsymbol{x}'|\boldsymbol{b}'+\boldsymbol{x}'|\boldsymbol{a}'+\boldsymbol{b}'+\boldsymbol{x}'$ can only be the same if $\boldsymbol{a} = \boldsymbol{a}'$, $\boldsymbol{b} = \boldsymbol{b}'$ and $\boldsymbol{x} = \boldsymbol{x}'$. Equality of the words implies

$$\boldsymbol{a}+\boldsymbol{x} = \boldsymbol{a}'+\boldsymbol{x}' \ , \ \boldsymbol{b}+\boldsymbol{x} = \boldsymbol{b}'+\boldsymbol{x}' \ , \ \boldsymbol{a}+\boldsymbol{b}+\boldsymbol{x} = \boldsymbol{a}'+\boldsymbol{b}'+\boldsymbol{x}'.$$

Adding the first two we obtain $\boldsymbol{a}+\boldsymbol{b} = \boldsymbol{a}'+\boldsymbol{b}'$, and combining with the third gives $\boldsymbol{x} = \boldsymbol{x}'$, then the first and second imply $\boldsymbol{a} = \boldsymbol{a}'$ and $\boldsymbol{b} = \boldsymbol{b}'$.

28. $w(\boldsymbol{x}+\boldsymbol{y}+\boldsymbol{z}) = \boldsymbol{w}((\boldsymbol{x}+\boldsymbol{y})+\boldsymbol{z})$
$= w(\boldsymbol{x}+\boldsymbol{y}) + w(\boldsymbol{z}) - 2w((\boldsymbol{x}+\boldsymbol{y}) \odot \boldsymbol{z})$
$= w(\boldsymbol{x}) + w(\boldsymbol{y}) + w(\boldsymbol{z}) - 2w(\boldsymbol{x} \odot \boldsymbol{y}) - 2w((\boldsymbol{x} \odot \boldsymbol{z}) + (\boldsymbol{y} \odot \boldsymbol{z}))$
$= w(\boldsymbol{x}) + w(\boldsymbol{y}) + w(\boldsymbol{z}) - 2w(\boldsymbol{x} \odot \boldsymbol{y}) - 2w(\boldsymbol{x} \odot \boldsymbol{z}) - 2w(\boldsymbol{y} \odot \boldsymbol{z}) + 4w((\boldsymbol{x} \odot \boldsymbol{z}) \odot (\boldsymbol{y} \odot \boldsymbol{z}))$.

29. Let $w(\boldsymbol{a}) = w_1$ and $w(\boldsymbol{b}) = w_2$, which we know are multiples of 4. Let p be the number of places in which $\boldsymbol{a}$, $\boldsymbol{b}$ are both 1, so $w(\boldsymbol{a} \odot \boldsymbol{b}) = p$. Hence there are $w_1 - p$ places where $\boldsymbol{a}$ is 1 and $\boldsymbol{b}$ is 0, and $w_2 - p$ places where $\boldsymbol{a}$ is 0 and $\boldsymbol{b}$ is 1. So

$$d(\boldsymbol{a},\ \boldsymbol{b}) \ = \ (w_1 - p) + (w_2 - p) = w_1 + w_2 - 2p.$$

Now for a linear code the distance and weight distributions are identical (Theorem 6.5), so $4|d(a,b)$, so $4|2p$ and p is therefore even.

30. From Theorem 6.19 it suffices to show that the dual code $C^{\perp}$, that is the one generated by H, is non-cyclic. This can be done by finding an

example of a codeword of $C^\perp$ whose first cyclic shift is not in $C^\perp$. One example is the first row of H, whose first cyclic shift is not expressible as a linear combination of the rows of H. Check this.

31.

$$(1+(q-1)x)^n A\left(\frac{1-x}{1+(q-1)x}\right) = (1+(q-1)x)^n \sum_{i=0}^{n} A_i\left(\frac{1-x}{1+(q-1)x}\right)^i$$

$$= \sum_{i=0}^{n} A_i(1-x)^i\,(1+(q-1)x)^{n-i}$$

32. From the expression for $B(x)$, the x^2 coefficient is

$$\left[\binom{n}{2}(q-1)^2 + (q-1)\binom{n}{2}\right] q^{-1} = \binom{n}{2}(q-1).$$

Hence the number of codewords of weight 2 is positive, so the code cannot have $d \geq 3$.

33. For $q = 2$ the expression for $B(x)$ simplifies to

$$B(x) = \frac{1}{2}\left[(1+x)^n + (1-x)^n\right] = \sum_{i=0}^{n} B_i x^i.$$

Expanding binomials $(1+x)^n$ and $(1-x)^n$ this gives $B_i = 0$ for all odd i, and for the even

$$i\ b_i = \binom{n}{i},$$

which is the number of words of weight i in Z_2^n.

Alternatively, the repetition code has only two words, $\mathbf{0}$ and $\mathbf{1} = 111\cdots1$. All words are orthogonal to $\mathbf{0}$ and a word is orthogonal to $\mathbf{1}$ if and only if it has even weight.

35. $X = A_w$ is just the definition of A_w. For each of the A_{w+1} codewords $\mathbf{c}$ of weight $w+1$, there are $w+1$ ways of changing this to a word $\boldsymbol{u}$ of weight w with $d(\mathbf{c},\ \boldsymbol{u}) = 1$ since just one of the $w+1$ non-zero symbols of $\mathbf{c}$ must be selected and changed to zero. Finally, for each of the A_{w-1} codewords of weight $w-1$, say $\mathbf{c}'$, there are $(n-w+1)(q-1)$ ways of changing it to a word $\boldsymbol{u}'$ of weight w with $d(\mathbf{c}',\ \boldsymbol{u}') = 1$ because one of its $n-w+1$ zero symbols must be selected and this symbol changed to one of the $q-1$ non-zero symbols.

36. Ham$(4,2)$ has $n = 15$, and for $q = 2$ the relation becomes

$$\binom{15}{w} = A_w + A_{w+1}(w+1) + A_{w-1}(16-w)$$

$A_1 = A_2 = 0$, so $A_3 = 35$, $A_4 = 105$, $A_5 = 168$

37. For $p = 0.01$, $2\sqrt{p(1-p)} = 2\sqrt{0.0099}$. For Ham (3, 2), $A_0 = 1, A_1 = A_2 = 0$ (for one-error correcting perfect codes), and by the method of Exercise 36 or direct use of the MacWilliams identity on the corresponding dual simplex code,
$A_3 = 7,\ A_4 = 7,\ A_5 = A_6 = 0,\ A_7 = 1$.
So $p(\text{error}) \leq 0.066$.
But for this code the received word is correctly decoded if and only if there are no errors or 1 error.

Hence $P(\text{error}) = 1 - (1-p)^7 - 7p(1-p)^6 = 0.002$.

38. Every odd weight codeword of C has its weight increased by 1 in C', and the even weight codewords have unchanged weight. Hence $A'_i = A_{i-1} + A_i$ for even i and $A'_i = 0$ for odd i.

39. The suggested codewords have weights 4, 5, 6, 5, 5, 5 respectively, and no pair of them are dependent, so the non-zero multiples of these codewords account for 24 codewords. But C has dimension 2 over Z_5 so has 25 codewords in all, so only **0** needs to be added. Multiplying a word by a non-zero constant does not change its weight, so for C $A_0 = 1,\ A_1 = A_2 = A_3 = 0,\ A_4 = 4,\ A_5 = 16,\ A_6 = 4$.

Applying the MacWilliams identity the dual code has

$A'_0 = 1,\ A'_1 = 0,\ A'_2 = 4,\ A'_3 = 64,\ A'_4 = 144,\ A'_5 = 248,\ A'_6 = 164$.

Chapter 7

1. $\deg(f+g) \leq \max\{\deg(f), \deg(g)\}$; $\deg(f-g) \leq \max\{\deg(f), \deg(g)\}$; $\deg(fg) = \deg(f) + \deg(g)$.

2. (a) Let $q(x) = q_2x^2 + q_1x + q_0$, $r(x) = r_3x^3 + r_2x^2 + r_1x + r_0$.

Then equating coefficients of $x^6, x^5, \cdots, x^0$ we obtain the equations (in Z_5):

$$\begin{aligned} 3 &= 4q_2,\ 2 = 4q_1 + 1q_2,\ 0 = 4q_0 + 1q_1 + 1q_2, \\ 4 &= 1q_0 + 1q_1 + 3q_2 + r_3,\ 0 = 1q_0 + 3q_1 + 1q_2 + r_2, \\ 2 &= 3q_0 + 1q_1 + r_1,\ 2 = 1q_0 + r_0. \end{aligned}$$

Solving these : $q_2 = 2, q_1 = 0, q_0 = 2, r_3 = 1, r_2 = 1, r_1 = 1, r_0 = 0$, so $q(x) = 2x^2 + 2$ and $r(x) = x^3 + x^2 + x$.

Using a similar method

for (b): $q(x) = 4$, $r(x) = 2x^5 + x^4 + 4x^3 + 3x^2 + 2x + 2$, and

for (c): $q(x) = x^5 + x^4 + 2x^3 + 2x^2 + 2x + 3$, $r(x) = 0$.

3.(a)

$$
\begin{array}{r|l}
 & x^2 \qquad + 1 \\
\cline{2-2}
x^4 + x^2 + x + 1 & x^6 \qquad + x^3 + x^2 + x \\
 & x^6 + x^4 + x^3 + x^2 \\
\cline{2-2}
 & \qquad x^4 \qquad + x \\
 & \qquad x^4 + x^2 + x + 1 \\
\cline{2-2}
 & \qquad\qquad x^2 \qquad + 1
\end{array}
$$

So $x^6 + x^3 + x^2 + x = (x^4 + x^2 + x + 1)(x^2 + 1) + (x^2 + 1)$

(b)

$$
\begin{array}{r|l}
 & 2x^3 \qquad + 2x + 1 \\
\cline{2-2}
2x^3 + x^2 + x + 1 & x^6 + 2x^5 \qquad\qquad + 2x + 1 \\
 & x^6 + 2x^5 + 2x^4 + 2x^3 \\
\cline{2-2}
 & \qquad x^4 + x^3 \qquad + 2x + 1 \\
 & \qquad x^4 + 2x^3 + 2x^2 + 2x \\
\cline{2-2}
 & \qquad\qquad 2x^3 + x^2 \qquad + 1 \\
 & \qquad\qquad 2x^3 + x^2 + x + 1 \\
\cline{2-2}
 & \qquad\qquad\qquad 2x
\end{array}
$$

So $x^6 + 2x^5 + 2x + 1 = (2x^3 + x^2 + x + 1)(2x^3 + 2x + 1) + 2x$

4. $g(-1) = g(1) =$ the total number of non-zero terms, modulo 2.

5. Let $\deg(f) = k$. Then

$$
\begin{aligned}
(x-\alpha)|f(x) &\Leftrightarrow f(x) = (x-\alpha)g(x) = (x-\alpha)(b_{k-1}x^{k-1} + \cdots + b_0) \\
&\Leftrightarrow f(x) = \beta(x-\alpha)(b_{k-1}\beta^{-1}x^{k-1} + \cdots + b_0\beta^{-1}) \\
&\Leftrightarrow \beta(x-\alpha)|f(x).
\end{aligned}
$$

6.(a) $3(x+4)^2$,

(b) $2(x^3 + x^2 + 2x + 1)$,

(c) $(x+6)(x+3)(x+2)$,

(d) $(x+1)(x^2 + x + 2)$,

(e) $2(x^2 + 5x + 2)$.

7. $f(0) = 0$, but $f(2)$, $f(3)$, $f(4)$ are clearly non-zero, while $f(1) = 2.3.4 + 4 \neq 0$. Hence $x|f(x)$, and on expanding the given expression for $f(x)$ we obtain
$f(x) = x(x^4 + x^3 + 2x^2 + 2x + 1)$, and we know that the quartic

factor has no linear factors. Hence the only possible candidate for further factorization is

$$x^4 + x^3 + 2x^2 + 2x + 1 = (x^2 + ax + b)(x^2 + a'x + b').$$

Equating coefficients, this gives

$a + a' = 1,\quad b + aa' + b' = 2\ ,\quad ab' + a'b = 2\ ,\quad bb' = 1.$

From the last of these equations $(b, b') = (1,1),\ (2,3),\ (3,2)$ or $(4,4)$, and it is easy to check that each of these possibilities makes the remaining three equations inconsistent. Hence $x^4+x^3+2x^2+2x+1$ is irreducible.

8. $f(x) = g(x)(x^4 + x^2 + 2x + 1) + (x^7 + x^6 + x^4 + x + 2)$

$g(x) = (x^7 + x^6 + x^4 + x + 2)(x + 2) + (x^4 + x + 1)$

$x^7 + x^6 + x^4 + x + 2 = (x^4 + x + 1)(x^3 + x^2) + (x^3 + 2x^2 + x + 2)$

$x^4 + x + 1 = (x^3 + 2x^2 + x + 2)(x + 1) + (x + 2)$

$x^3 + 2x^2 + x + 2 = (x + 2)(x^2 + 1) + 0$

So gcd $(f(x), g(x)) = x + 2.$

9.(a) Working mod $x^3 + 2x$, $\quad x^3 + 2x \equiv 0.$

$$\begin{array}{llllllll} \text{Hence} & x^3 & \equiv & -2x & \equiv & x & & \\ \Rightarrow & x^4 & = & x(x^3) & \equiv & xx & = & x^2 \\ \Rightarrow & x^5 & = & x(x^4) & \equiv & x^3 & \equiv & x \\ \Rightarrow & x^6 & = & x(x^5) & \equiv & x^2 & & \\ \Rightarrow & x^7 & = & x(x^6) & \equiv & x^3 & \equiv & x\ . \end{array}$$

So $x^7+2x^6+2x^4+x^3+x^2+2 \equiv x+2x^2+2x^2+x+x^2+2 \equiv 2x^2+2x+2$

(b) $x^3 \equiv -2x \equiv 3x \Rightarrow x^4 \equiv 3x^2 \Rightarrow x^5 \equiv 4x \Rightarrow x^5 \equiv 4x^2$

$\Rightarrow x^6 \equiv 2x \Rightarrow x^7 \equiv 2x^2$

So $x^7+2x^6+2x^4+x^3+x^2+2 \equiv 2x^2+4x+x^2+3x+x^2+2 \equiv 4x^2+2x+2.$

10. They are all rings. (i), (ii) (iii) have unity

$$\begin{bmatrix} 1 & 0 & 0 \\ 0 & 1 & 0 \\ 0 & 0 & 1 \end{bmatrix}, \quad \begin{bmatrix} 1 & 0 \\ 0 & 1 \end{bmatrix}, \quad \begin{bmatrix} 0 & 0 \\ 0 & 1 \end{bmatrix}$$

respectively.

(iv) has no unity.

Only (iv) is a commutative ring.

11. Let J be the set of all multiples of $\alpha m(x)$ by polynomials in $F[X]$, so we have to prove $I = J$.

$$\begin{aligned} f \in I \Rightarrow \quad f(x) &= m(x)g(x) \text{ (for some } g \in F[X]) = m(x)\left[a_n x^n + \cdots\right] \\ &= \alpha m(x)\left[\alpha^{-1} a_n x^n + \cdots\right] \text{ where } n = \text{ deg } (g). \\ \Rightarrow \quad f(x) &\in J. \end{aligned}$$

Similarly it can be shown that $f \in J \Rightarrow f \in I$.

Now let $m(x), m'(x)$ be any two generators of minimal degree d for I. By dividing each by their leading coefficients we obtain two monic generators $M(x)$, $M'(x)$.

Now $M(x) - M'(x) \in I$ and $\deg(M - M') < d$.

Hence $M(x) = M'(x)$ so $m'(x)$ is a constant multiple of $m(x)$.

12. The subset is clearly a subring of $F[X, Y]$ and the ideal properties are easily checked. Suppose it is a principal ideal, and let $g \in F[X,\ Y]$ be a generator. Clearly $x \in I$ (take $s(x, y) = 1$, $t(x, y) = 0$), and similarly $y \in I$. Hence x is a multiple of $g(x, y)$ so $g(x, y)$ cannot contain any terms involving y. Similarly it cannot have any x-terms. Hence $g(x, y)$ is a constant polynomial k. But $g \in I$, so $k = xs(x, y) + yt(x, y)$ for some s, t in $F[X,\ Y]$. This is not possible unless $k = 0$. But then $I = \{0\}$ which contradicts the fact that $x \in I$.

Hence I has no single generator, so is not principal.

Chapter 8

1. $2x$.

3. In R $1 + x^2 = (1 + x)^2$ and $1 + x = x^2(1 + x^2)$, so $\langle 1 + x^2 \rangle \subseteq \langle 1 + x \rangle$ and $\langle 1 + x \rangle \subseteq \langle 1 + x^2 \rangle$ or, if you don't spot this simply multiply $1 + x$ by each of the 8 members of R and reduce mod $x^3 - 1$. Do the same for $1 + x^2$ observe that you get the same set of members of R.

4. Consider the equation $\lambda_1 r_1 + \cdots + \lambda_k r_k = 0$ where r_1 is the ith row of G. Since $g_0 \neq 0$, this equation gives successively $\lambda_1 = 0$, $\lambda_2 = 0, \ldots$.

5. (i) Not linear, w cannot be cyclic.
   (ii) Linear but not cyclic. Switch positions 2 and 3 to obtain the cyclic code $\{0000, 1212, 2121\}$, so the original code is linearly equivalent to a cyclic code.
   (iii) Cyclic.
   (iv) Cyclic.
   (v) Not linear, e.g. 0111 and 1110 $\in C$ but $0111 + 1110 = 1221 \notin C$.
   (vi) Cyclic.

6.(a) $x^{n-1} + x^{n-2} + \cdots + x^2 + x + 1$.

(b) Every multiple of $g(x)$ is a multiple of $(x-1)$. If $h(x)$ has k non-zero terms, and $(x-1)h(x)$ is multiplied out it gives a polynomial with $2k$ non-zero terms, some of which may cancel, but only in pairs. Hence the corresponding codeword has even weight.

(c) $q(x)$ has n (odd) non-zero terms. Hence $(x-1) \not| q(x)$ by the argument used in (b).

(d) Let E be the even subcode of C. Every word in $\langle (x-1)g(x) \rangle$ is in C and has even weight. Hence $\langle (x-1)g(x) \rangle \subseteq E$. Conversely, let $e(x) \in E$. Then $e(x) = g(x)p(x)$ for some polynomial p. $(x-1)|e(x)$ and $(x-1) \not| g(x)$, so $(x-1)|p(x)$, and $e(x) \in \langle (x-1)g(x) \rangle$, so $E \subseteq \langle (x-1)g(x) \rangle$, and we have $E = \langle (x-1)g(x) \rangle$. $E \subset C$ so $\dim(E) < \dim C$. Corollary to Theorem 8.4 then implies that the generator polynomial of E has higher degree than that of C. Hence no multiple of $(x-1)g(x)$ has smaller degree, so this polynomial is *the* generator of E.

Finally $g(x)|(x-1)q(x)$ and $\gcd(g(x), x-1) = 1$ so $g(x)|q(x)$. Hence $q(x) \in \langle g(x) \rangle = C$, and $q(x)$ is the word $111\ldots1$.

7. No. $x^3+2x^2+2 = (x+1)(x^2+x+2)$ over Z_3 so x^3+2x^2+2 is reducible.

8. We require divisions of $x^{21}-1$ over Z_2 with degree $21-9=12$. From the factor table, writing the given factorization as $x^{21}-1 = h_1h_2h_3h_3'h_6h_6'$ where the subscript denotes the degree, those of degree 12 are h_6h_6', $h_6h_3h_3'$, $h_6'h_3h_3'$, $h_6h_3h_2h_1$, $h_6h_3'h_2h_1$, $h_6'h_3h_2h_1$ and $h_6'h_3'h_2h_1$. The codes are those having these polynomials as their generators.

9. 112110 corresponds to $1+x+2x^2+x^3+x^4$ which factorizes into inreducibles over Z_3 as $(1+x)(2+2x+x^3)$. x^6-1 factorizes into $(1+x)^3(2+x)^3$, so the generator $(1+x)^a(2+x)^b$ $(a \leq 3, b \leq 3)$ of the code must satisfy $(1+x)(2+2x+x^3) \equiv \lambda(x)(1+x)^a(2+x)^b \bmod x^6-1$. An argument similar to that already illustrated gives 1 and $1+x$ as the only possible generators, $(1+x)$ giving the smaller code. $1+x$ corresponds to a word of weight 2 so $d(c) \leq 2$. If $d(c)=1$, then there must be some codeword x^k, but this would imply $x^k \equiv \mu(x)(1+x) \bmod x^6-1$. This is impossible, so $d(c)=2$.

10. $x^3 = g(x)+1+x^2$, $x^4 = (1+x)g(x)+1+x+x^2$, $x^5 = (1+x+x^2)g(x)+1+x$, $x^6 = (x+x^2+x^3)g(x)+x+x^2$.

So

$$G = \begin{bmatrix} 1 & 0 & 1 & 1 & 0 & 0 & 0 \\ 1 & 1 & 1 & 0 & 1 & 0 & 0 \\ 1 & 1 & 0 & 0 & 0 & 1 & 0 \\ 0 & 1 & 1 & 0 & 0 & 0 & 1 \end{bmatrix}.$$

11. $n=15$, $k=15-6=9$.

$$\begin{aligned} x^{n-k}m(x) &= x^6(x+x^4+x^8) = x^7+x^{10}+x^{14} \\ x^{n-k}m(x) &= g(x)(x^8+x^7+x^2+x)+x^4+x^2+1 = g(x)q(x)+r(x). \end{aligned}$$

So encoding of m is

$$\begin{aligned} q(x)g(x) &= x^{n-k}m(x) - r(x) \\ &= x^{14} + x^{10} + x^7 + x^4 + x^2 + 1 \\ &= 101010010010001. \end{aligned}$$

12. 110001.

13. $s(a)$, $s(a^1)$, …, $s(a^{15})$

$$= 110001, 111111, 111000, 011100, 001110, 000111, 100100, 010010,$$
$$001001, 100011, 110110, 011011, 101010, 010101, 101101, 110001.$$

14. $$\begin{aligned} \text{syn}(e) &= \text{syn}(r) = \text{ remainder on dividing } r(x) \text{ by } g(x) \\ &= 2x^4 + 2x^3 + 2x^2 + 2x, \end{aligned}$$

which corresponds to the word 02222.

By Theorem 8.10, $\text{syn}(r') = 20010$ which has weight 2, so

$$e' = 20010000000 \quad \text{and} \quad e = 00100000002.$$

Hence transmitted word $= r - e = 20021020110$. (It is worth checking that this is a codeword by evaluating its syndrome.)

The weight 2 errors which are not 'trapped' are those without a cyclic run of 6 zeros. There are $11 \times 2^2 = 44$ such errors, and $2^2 \times \binom{11}{2}$ weight 2 words. Hence the proportion is $\frac{1}{5}$.

15. $x^6(2x^{-6} + 3x^{-5} + x^{-3} + 4) = 4x^6 + x^3 + 3x + 2$

The sequences of coefficients in $p(x)$ and $x^6p(x^{-1})$ are $(2, 3, 0, 1, 0, 0, 4)$ and $(4, 0, 0, 1, 0, 3, 2)$.

16. The codes of Ham(3, 2) are all $[7, 4]$ codes. If C is a cyclic member of this family with generator $g(x)$, then $g(x)$ must be a divisor of $x^7 - 1$ over Z_2 with degree 3. Since $x^7 - 1 = (x + 1)(x^3 + x^2 + 1)(x^3 + x + 1)$, $g(x)$ must be one of the cubic divisors. The second one gives $h(x) = (x + 1)(x^3 + x^2 + 1) = x^4 + x^2 + x + 1$, $\bar{h}(x) = x^4 + x^3 + x^2 + 1$, so H is the matrix of section 6.7. [The other cubic divisor gives a different, but equivalent, cyclic code.]

17. $$\begin{aligned} \bar{h}(x)|g(x) &\Leftrightarrow \text{ every multiple of } g(x) \text{ is a multiple of } \bar{h}(x) \\ &\Leftrightarrow C \subseteq \langle \bar{h}(x) \rangle = C^{\perp}. \end{aligned}$$

From the table of factorizations of $x^n - 1$ over Z_2 we see that $1 + x + x^4 | x^{15} - 1$ and $1 + x^3 + x^4 | x^{15} - 1$. But $\overline{1 + x + x^4} = x^4 + x^3 + 1$, so if $g(x) = (1+x)(1+x+x^2)(1+x+x^2+x^3+x^4)(1+x^3+x^4)$, $h(x) = 1+x+x^4$, then $\bar{h}(x)|g(x)$, so $\langle g(x) \rangle$ is self-orthogonal.

18. From the table of factorizations $g(x) = (1 + x)(1 + x + x^4)$ satisfies the conditions of Theorem 8.13 so $\langle g \rangle$ is a suitable code.

19. If the columns of G are written in reverse order, and the rows of the result are then written in reverse order, the net result is G'.

20. If $j = k$ consider instead the cancelling pairs ij, kl; ki, lj. That is ij, jl; ji, lj and these are distinct unless $i = l$. But if $i = l$ we have $e_i - e_j \equiv e_j - e_i \bmod p$, so $2e_i \equiv 2e_j \bmod p$, which implies $e_i = e_j (i = j)$ because e_i and e_j are both between 0 and $p - 1$.

21. Over Z_2, $x^{23} - 1 = (x-1)(1+x^2+x^4+x^5+x^6+x^{10}+x^{11})(1+x++x^5+x^6+x^7+x^9+x^{11})$ which is of the form $(x-1)g(x)g^{\text{rev}}(x)$. Hence $\langle g \rangle$ is a cyclic binary code with length 23 and dimension $23 - 11 = 12$, and from Theorem 8.14 its minimum weight w is ≤ 7 since g has 7 terms. w cannot be even since it would have to be a multiple of 4 other than 4 itself ($p = 23 \neq 7$). Hence w satisfies $w^2 - w \geq p - 1 = 22$. Considering the quadratic graph $w^2 - w - 22 = 0$ for $w > 0$ shows that $w \geq 7$, so $w = 7$. Finally $\langle g \rangle$ is perfect by checking the sphere-packing bound for these parameters. There is only one (up to equivalence) binary perfect [23, 12, 7] code, so $\langle g \rangle$ is equivalent to the Golay code.

22. $D \neq C$ so there is a word $a \in C \backslash D$. Then all words of the coset $a + D$ are in C because for each $d \in D$, $a + d \in C$ by linearity of C. Hence C consists of complete cosets of D and since $|C| = 3|D|$, C is the union of three cosets of D.

Chapter 9

1. Since the codes are binary only closure under + needs to be checked. Let $\boldsymbol{x}, \boldsymbol{y} \in C_1 * C_2$. Then $\boldsymbol{x} = \boldsymbol{u}|\boldsymbol{u}+\boldsymbol{v}, \boldsymbol{y} = \boldsymbol{u}'|\boldsymbol{u}'+\boldsymbol{v}'$ for some $\boldsymbol{u}, \boldsymbol{u}' \in C_1$ and some $\boldsymbol{v}, \boldsymbol{v}' \in C_2$.

   Then $\boldsymbol{x} + \boldsymbol{y} = \boldsymbol{u} + \boldsymbol{u}'|\boldsymbol{u} + \boldsymbol{u}' + \boldsymbol{v} + \boldsymbol{v}'$ which clearly is in $C_1 * C_2$ since $\boldsymbol{u} + \boldsymbol{u}' \in C_1$ and $\boldsymbol{v} + \boldsymbol{v}' \in C_2$.

2. Suppose $\boldsymbol{u}|\boldsymbol{u} + \boldsymbol{v} = \boldsymbol{u}'|\boldsymbol{u}' + \boldsymbol{v}'$, then by comparing the first n places $\boldsymbol{u} = \boldsymbol{u}'$, and from the remaining n places $\boldsymbol{u} + \boldsymbol{v} = \boldsymbol{u}' + \boldsymbol{v}'$ so $\boldsymbol{v} = \boldsymbol{v}'$.

3. Let $\boldsymbol{c} = \boldsymbol{u}|\boldsymbol{u} + \boldsymbol{v} \in C_1 * C_2$. Then $\boldsymbol{c} = \boldsymbol{u}|\boldsymbol{u} + \boldsymbol{o}|\boldsymbol{v}$ and any word of the form $\boldsymbol{u}|\boldsymbol{u}$ is a linear combination of rows of $[G_1 \;\; G_1]$, and those of form $\boldsymbol{o}|\boldsymbol{v}$ are linear combinations of rows of $[O \;\; G_2]$. Hence the rows of the given matrix generate $C_1 * C_2$ and it just remains to prove the rows are independent.

   Now the matrix has $\dim C_1 + \dim C_2$ rows, so they must span a subspace of dimension $\leq \dim C_1 + \dim C_2$, with equality if and only if the rows are independent. But $\dim C_1 + \dim C_2$ *is* the dimension of $C_1 * C_2$ by part (ii) of the theorem, so the rows must be independent.

4.(a) Ham (3,2) is a [7, 4, 3] code. Hence D has parameters $n = 14, k = \dim C + \dim C = 8, d = \min(2 \times 3, 3) = 3$.

(b) The Hamming upper bound for D is

$$M \leq 2^{14} \left[\sum_{i=0}^{1} \binom{14}{i} \right]^{-1} = \frac{2^{14}}{15} ,$$

but we know $M = 2^8$ so D is not perfect.

(c) The simplest reason is that if all words of weight 2 were coset leaders then D would be 2-error correcting, and we know it isn't because d is only 3.

Another reason is that there are $\binom{14}{2} = 91$ words of weight 2, 14 of weight 1 and 1 of weight 0, giving a total of 106 words of weight at most 2. But the number of cosets is

$$\frac{|Z_2^{14}|}{|D|} = \frac{2^{14}}{2^8} = 2^6 = 64 \quad \text{– not enough!}$$

(d) A p.c. matrix for C is

$$\begin{bmatrix} 1 & 0 & 0 & 0 & 1 & 1 & 1 \\ 0 & 1 & 0 & 1 & 0 & 1 & 1 \\ 0 & 0 & 1 & 1 & 1 & 0 & 1 \end{bmatrix},$$

so a corresponding generator is

$$G = \begin{bmatrix} 0 & 1 & 1 & 1 & 0 & 0 & 0 \\ 1 & 0 & 1 & 0 & 1 & 0 & 0 \\ 1 & 1 & 0 & 0 & 0 & 1 & 0 \\ 1 & 1 & 1 & 0 & 0 & 0 & 1 \end{bmatrix}.$$

Using Exercise 3,

$$\begin{bmatrix} G & G \\ 0 & G \end{bmatrix}$$

is a generator of D,
and this is row equivalent to

$$\begin{bmatrix} G & 0 \\ 0 & G \end{bmatrix}$$

(by doing the row operations
$R_i \to R_i + R_{i+4}$ for $i = 1, 2, 3, 4$). This matrix now has the 8 unit words as 8 of its 14 columns, so by using the algorithm given in Chapter 5, a p.c. matrix for D is

$$H = \begin{bmatrix} 1 & 0 & 0 & 0 & 1 & 1 & 1 & 0 & 0 & 0 & 0 & 0 & 0 & 0 \\ 0 & 1 & 0 & 1 & 0 & 1 & 1 & 0 & 0 & 0 & 0 & 0 & 0 & 0 \\ 0 & 0 & 1 & 1 & 1 & 0 & 1 & 0 & 0 & 0 & 0 & 0 & 0 & 0 \\ 0 & 0 & 0 & 0 & 0 & 0 & 0 & 1 & 0 & 0 & 0 & 1 & 1 & 1 \\ 0 & 0 & 0 & 0 & 0 & 0 & 0 & 0 & 1 & 0 & 1 & 0 & 1 & 1 \\ 0 & 0 & 0 & 0 & 0 & 0 & 0 & 0 & 0 & 1 & 1 & 1 & 0 & 1 \end{bmatrix}$$

Since D is 1-error correcting the syndromes of all error patterns of weight 1 are distinct. They are just the columns of H.

For the errors of weight 2, syn(11000000000000) = 110000, which is the syndrome of the weight 1 error pattern 00000100000000, so cannot be correctly decoded. But syn (10000000010000) = 100001 which is not the syndrome of any weight 1 error pattern, so this error pattern can be made a coset leader, and will therefore be correctly decoded.

5. I am too lazy to write them all out, but there are 2^7 words, all of length 8.
6. Clearly from Definition 9.2 ($RM(o,m)$ and $RM(m,m)$ are linear and of length 2^m for all m. For general r, m, clause 3 of the definition, Theorem 9.1(i) and induction on m will show that all the Reed–Muller codes are linear, and that their lengths are 2^m.
7. First prove that for all $m \geq 1, RM(m-1,m)$ is $E(2^m)$, the set of all even weight words of length 2^m. For $m = 1$ this is clearly true, so suppose it is true for all $m < k$.

   Let
   $$\begin{aligned} \boldsymbol{c} \in RM(k-1,k) &= RM(k-1,k-1) * RM(k-2,k-1) \\ &= Z_2^{2^{k-1}} * E(2^{2^{k-1}}) \end{aligned}$$
   by the induction hypothesis.

   So $\boldsymbol{c} = \boldsymbol{u}|\boldsymbol{u}+\boldsymbol{v}$ for some $\boldsymbol{u} \in Z_2^{2^{k-1}}, \boldsymbol{v} \in E(2^{k-1})$.

   Now using $w(\boldsymbol{u}+\boldsymbol{v}) = w(\boldsymbol{u}) + w(\boldsymbol{v}) - 2w(\boldsymbol{u} \odot \boldsymbol{v})$ we see that $\boldsymbol{u}$ and $\boldsymbol{u}+\boldsymbol{v}$ have the same parity, so $\boldsymbol{c}$ must be in $E(2^k)$.

   Conversely, if $\boldsymbol{c}$ is any word of $E(2^k)$, write $\boldsymbol{c}$ as $\boldsymbol{x}|\boldsymbol{y}$ where $\boldsymbol{x}, \boldsymbol{y}$ have length 2^{k-1} and must have the same parity. But $\boldsymbol{y} = \boldsymbol{x} + (\boldsymbol{x}+\boldsymbol{y})$ so $\boldsymbol{x}+\boldsymbol{y}$ must have even weight.

   So $\boldsymbol{c} \in RM(k-1,k)$.

   Now any word of $Z_2^{2^m}$ is either of even weight, or if not, can be obtained by adding $000\ldots01$ to an even weight word. Hence $Z_2^{2^m}$ is generated by the even weight words and $000\ldots01$, and $G(m,m) = Z_2^{2^m}$ has a generator matrix
   $$\begin{bmatrix} G(m-1, & m) \\ 000 & \ldots\ 01 \end{bmatrix} \qquad \ldots(1).$$
   At the other extreme $G(o,m)$ is just the repetition code of length 2^m so its generator matrix is $[111\ \ldots\ 1]$ ...(2).

   For all other relevant values of r and m, that is $o < r < m$, Exercise 3 gives the result:
   $$G(r,m) = \begin{bmatrix} G(r,m-1), & G(r,m-1) \\ 0, & G(r-1,m-1) \end{bmatrix} \qquad \ldots(3),$$
   and (1), (2), (3) provide a full definition of $G(r,m)$ for $o \leq r \leq m$.

8. The theorem is clearly true for $r = m$ and for $r = 0$.

   Now suppose $d(RM(r,k)) = 2^{k-r}$ for some k and all $r < k$. $RM(r, k+1) = RM(r,k) * RM(r-1,k)$, so by Theorem 9.1(iii),

$$\begin{aligned} d(RM(r,k+1)) &= \min\{2 \times d(RM(r,k)), d(RM(r-1,k))\} \\ &= \min\{2 \times 2^{k-r}, 2^{k-(r-1)}\} \\ &= 2^{(k+1)-r} \end{aligned}$$

   so by induction the result follows.

9.

$$G(2,3) = \begin{bmatrix} 1 & 1 & 1 & 1 & 1 & 1 & 1 & 1 \\ 0 & 1 & 0 & 1 & 0 & 1 & 0 & 1 \\ 0 & 0 & 1 & 1 & 0 & 0 & 1 & 1 \\ 0 & 0 & 0 & 1 & 0 & 0 & 0 & 1 \\ 0 & 0 & 0 & 0 & 1 & 1 & 1 & 1 \\ 0 & 0 & 0 & 0 & 0 & 1 & 0 & 1 \\ 0 & 0 & 0 & 0 & 0 & 0 & 1 & 1 \end{bmatrix}$$

$$G(2,4) = \begin{bmatrix} G(2,3) & & & & G(2,3) & & & \\ 0\ldots0 & 1 & 1 & 1 & 1 & 1 & 1 & 1 & 1 \\ \vdots & 0 & 1 & 0 & 1 & 0 & 1 & 0 & 1 \\ \vdots & 0 & 0 & 1 & 1 & 0 & 0 & 1 & 1 \\ 0\ldots0 & 0 & 0 & 0 & 0 & 1 & 1 & 1 & 1 \end{bmatrix}$$

10. For a 7-error correcting code we require $d \geq 15$. But for Reed–Muller codes $d = 2^{m-r}$ so we require $m - r \geq 4 \quad \ldots(1)$. $M = 2^{f(r,m)}$ so for $M \geq 1000$ we require $f(r,m) \geq 10$.

$$\text{i.e.} \qquad \sum_{i=0}^{r} \binom{m}{i} \geq 10 \qquad \ldots(2)$$

   Finally, to get a minimal length code we take the smallest value of m consistent with (1) and (2). Trying $m = 4, r = 0$; $m = 5, r = 1$; $m = 6, r = 2$ we find that $RM(2,6)$ is the shortest code satisfying the requirements.

11.

$$\begin{aligned} \bar{x}\bar{y}\bar{z} + x\bar{y}z &= (1 \oplus x)(1 \oplus y)(1 \oplus z) + x(1 \oplus y)z && \text{from 4} \\ &= (1 \oplus x)(1 \oplus y)(1 \oplus z) \oplus x(1 \oplus y)z \\ & \qquad \text{from 2 and the fact that } x(1 \oplus x) = 0, \\ & \qquad \text{and } o.a = 0 \text{ for all } a, \text{ and 5} \\ &= 1 \oplus x \oplus y \oplus z \oplus xy \oplus xz \oplus yz \oplus xyz \oplus xz \oplus xyz \\ & \qquad \text{by repeated use of 7 and 1} \\ &= 1 \oplus x \oplus y \oplus z \oplus xy \oplus yz && \text{from 6 and 5.} \end{aligned}$$

12.

$$
\begin{array}{rcl}
 & xyz + xy\bar{z} & + x\bar{y}\bar{z} + \bar{x}\bar{y}z \\
= & xy(z + \bar{z}) & + \bar{y}(x\bar{z} + \bar{x}z) \\
= & xy & + (1 \oplus y)(x \oplus z) \\
= & xy & \oplus (1 \oplus y)(x \oplus z) \oplus xy(1 \oplus y)(x \oplus z) \\
= & xy & \oplus (1 \oplus y)(x \oplus z)(1 \oplus xy) \\
= & xy & \oplus (x \oplus z \oplus yx \oplus yz)(1 \oplus xy) \\
= & xy & \oplus x \oplus z \oplus yx \oplus yz \oplus xy \oplus xyz \oplus xy \oplus xyz \\
= & x & \oplus z \oplus yz
\end{array}
$$

13. The words of $C * C$ have the form $\boldsymbol{u}|\boldsymbol{u}+\boldsymbol{v}$ where $\boldsymbol{u}, \boldsymbol{v}$ are any words of C. (1)

    But if $\mathbf{c}$ is any word of C it can be written as $\boldsymbol{u} + (\boldsymbol{u} + \mathbf{c})$, and by the linearity of C, $\boldsymbol{u}+\mathbf{c}$ is also in C, so the word $\boldsymbol{u}|\mathbf{c}$ is of the form (1) above.

    Hence $C = \{\boldsymbol{u}|\mathbf{c} : \boldsymbol{u} \in C, \mathbf{c} \in C\}$.

14. Let $f(x_1, \ldots, x_m), g(x_1, \ldots, x_m) \in BF(r, m)$. Then thinking of $\boldsymbol{f}$ and $\boldsymbol{g}$ as being given by their binary columns of length 2^m, clearly $0.\boldsymbol{f}$ and $1.\boldsymbol{f} \in BF(r, m)$.

    Also, $\boldsymbol{f} \oplus \boldsymbol{g}$ is another function of m variables and its degree is necessarily $\leq r$ since it is $\leq \max\{\deg \boldsymbol{f}, \deg \boldsymbol{g}\}$.

15. $BF(m, m)$ is the set of *all* Boolean functions of m variables, and for $r > m$ $BF(r, m) = BF(m, m)$ since no term in the $\oplus$ expansion of a function of m variables can have degree $> m$ because all repeated occurrences of a variable can be suppressed since $a^n = a$.

16. Theorem 9.2: Each term of the $\oplus$ expansion of a Boolean function corresponds to the subset of the m variables which appear in it. So the subset corresponding to a term in $BF(r, m)$ has at most r members. Hence the total number of terms which a member of $BF(r, m)$ can contain is

$$\sum_{i=0}^{r} \binom{m}{i} = f(r, m).$$

    The particular function in $BF(r, m)$ is determined by which subsets of these $f(r, m)$ terms actually appear in the $\oplus$ expansion of the function, and a set with $f(r, m)$ members has precisely $2^{f(r,m)}$ subsets.

    Theorem 9.4: $BF(r - 1, m)$ is clearly a subset of $BF(r, m)$!.

17. The basis step requires proof that each word of $RM(r, 1)$ is orthogonal to each word of $RM(-r, 1)$. This only makes sense for $r = 0$, in which case the matrix is $[\ 11\ ]$ and clearly 11 is self-orthogonal.

18. From Theorem 9.2,

$$\begin{aligned}
\dim(RM(m-r-1,m)) &= f(m-r-1,m)\\
&= \sum_{i=0}^{m-r-1} \binom{m}{i} = \sum_{i=r+1}^{m} \binom{m}{i}\\
&\quad \text{by the symmetry property of binomial coefficients}\\
&= \sum_{i=0}^{m}\binom{m}{i} - \sum_{i=0}^{r}\binom{m}{i}\\
&= 2^m - \dim(RM(r,m)) = \dim[RM(r,m)]^{\perp}.
\end{aligned}$$

References

[1] Hill, R. (1986) *A First Course in Coding Theory*, Oxford University Press.

[2] Wesley Peterson, W. (1962) Error-correcting codes. *Scientific American*, **206(2)**, pp 96–108.

[3] McEliece, R.J. (1985) The reliability of computer memories. *Scientific American*, **252(1)**, pp 88–95.

[4] Baylis, J. and Haggarty, R. (1988) *Alice in Numberland*, Macmillan.

[5] Burton, D. (1997) *Elementary Number Theory*, 3rd edn, McGraw Hill.

[6] Allenby, R.B.J.T. and Redfern, E.J. (1989) *Introduction to Number Theory with Computing*, Edward Arnold.

[7] Rosen, K.H. (1992) *Elementary Number Theory and its Applications*, 3rd edn, Addison-Wesley.

[8] Mills, J.T.S. (1996) Another family tree of Pythagorean triples. *The Mathematical Gazette*, **80**, 489, pp 545–548.

[9] Pompili, F. (1996) Evolution of finite sequences of integers. *The Mathematical Gazette*, **80**, 488, pp 322–332.

[10] Körver, W.H.F.J. (1996) Matches and coins, an old game with new rules. *The Mathematical Gazette*, **80**, 487, pp 243–244.

[11] MacWilliams, F.J. and Sloane, N.J.A. (1993) *The Theory of Error-correcting Codes*, North-Holland.

[12] Wesley Peterson, W. and Weldon, E.J. Jr (1991) *Error-correcting Codes*, MIT Press.

[13] Allenby, R.B.J.T. (1995) *Linear Algebra*, Edward Arnold.

[14] Cameron, P.J. and van Lint, J.H. (1991) *Designs, graphs, codes and their links*, Cambridge University Press.

[15] Bryant, V. (1993) *Aspects of Combinatorics*, Cambridge University Press.

[16] Pretzel, O. (1992) *Error-correcting Codes and Finite Fields*, Oxford University Press.

[17] Vanstone, S.A. and van Oorschot, P.C. (1992) *An Introduction to Error-correcting Codes with Applications*, Kluwer.

Index